Philosophy and Geography I:

SPACE, PLACE, AND ENVIRONMENTAL ETHICS

Philosophy and Geography
A Peer Reviewed Annual
Editors: Andrew Light, Department of Philosophy, University of Montana, and Jonathan M. Smith, Department of Geography, Texas A&M University
Sponsored by the Society for Philosophy and Geography

Volume I: Space, Place, and Environmental Ethics

Volume II: Public Space, forthcoming, October 1997.

Volume III: The Meaning of Place, submission deadline: September 15, 1997.

Volume IV: Aesthetics of Everyday Life, submission deadline: September 15, 1998.

See page 282 for submission guidelines.

Editorial Board
Albert Borgmann, philosophy, University of Montana
Augustin Berque, École des Hautes Études en Sciences Sociales, Paris
J. Baird Callicott, philosophy, University of North Texas
Edward Casey, philosophy, SUNY Stony Brook
Denis Cosgrove, geography, Royal Holloway, University of London
Arthur Danto, philosophy, Columbia University
James Duncan, geography, Cambridge University
Avner De-Shalit, political science, Hebrew University of Jerusalem
J. Nicholas Entrikin, geography, UCLA
Andrew Feenberg, philosophy, San Diego State University
Mark Gottdiner, sociology, University of Buffalo
Derek Gregory, geography, University of British Columbia
David Harvey, geography, Johns Hopkins University
Kathleen Marie Higgins, philosophy, University of Texas, Austin
Bernd Magnus, philosophy and humanities, University of California, Riverside
Thomas McCarthy, philosophy, Northwestern University
Bryan Norton, School of Public Policy, Georgia Institute of Technology
Carole Pateman, political science, UCLA
John Pickles, geography, University of Kentucky
Moishe Postone, history, University of Chicago
Juval Portugali, geography, Tel Aviv University
David Seamon, architecture, Kansas State University
Neil Smith, geography, Rutgers University
James Wescoat, Jr., geography, University of Colorado
Iris Marion Young, Graduate School of Public and International Affairs, University of Pittsburgh

Associate Editors: Yoko Arisaka, University of San Francisco; Jean-Marc Besse, Collège International de Philosophie à Paris; Edward Dimendberg, University of California Press; Thomas Heyd, University of Victoria; Eric Katz, New Jersey Institute of Technology; William Lynn, University of Minnesota; Jonathan Maskit, Denison University; James Proctor, University of California, Santa Barbara; Rupert Read, University of Manchester

Philosophy and Geography I:

SPACE, PLACE, AND ENVIRONMENTAL ETHICS

Edited by
ANDREW LIGHT
JONATHAN M. SMITH

ROWMAN & LITTLEFIELD PUBLISHERS, INC.
Lanham • Boulder • New York • London

ROWMAN & LITTLEFIELD PUBLISHERS, INC.

Published in the United States of America
by Rowman & Littlefield Publishers, Inc.
4720 Boston Way, Lanham, Maryland 20706

3 Henrietta Street
London WC2E 8LU, England

Copyright © 1997 by Rowman & Littlefield Publishers, Inc.

All rights reserved. No part of this publication may be
reproduced, stored in a retrieval system, or transmitted
in any form or by any means, electronic, mechanical,
photocopying, recording, or otherwise, without the prior
permission of the publisher.

British Cataloging in Publication Information Available

ISSN 1090-3771

ISBN 0–8476–8220-X (cloth : alk. paper)
ISBN 0–8476 –8221-8 (pbk. : alk. paper)

Printed in the United States of America

∞™ The paper used in this publication meets the minimum requirements of American National Standard for Information Sciences—Permanence of Paper for Printed Library Materials, ANSI Z39.48-1984.

Contents

List of Illustrations	ix
Acknowledgments	xi
Introduction: Geography, Philosophy, and the Environment *Andrew Light and Jonathan M. Smith*	1
On the Ethical Determination of Geography: A Kantian Prolegomenon *Robert Burch*	15
Nature's Presence: Reflections on Healing and Domination *Eric Katz*	49
The Takings Clause and the Meanings of Land *Zev Trachtenberg*	63
Muslim Contributions to Geography and Environmental Ethics: The Challenges of Comparison and Pluralism *James L. Wescoat, Jr.*	91
The Dialectical Social Geography of Elisée Reclus *John Clark*	117
The Maintenance of Natural Capital: Motivations and Methods *Clive L. Spash and Anthony M. H. Clayton*	143
Wilderness Management *Roger Paden*	175
Mead and Heidegger: Exploring the Ethics and Theory of Space, Place, and the Environment *Eliza Steelwater*	189

Critical Reflections on Biocentric Environmental Ethics: Is It
an Alternative to Anthropocentrism? 209
  Roger King

Ecology, Modernity, and the Intellectual Legacy of the
Frankfurt School 231
  Matthew Gandy

Critical Questions in Environmental Philosophy 255
  Annie L. Booth

Index 275

About the Contributors 281

Illustrations

Figures

5.1. Overlapping Terrain I	92
5.2. Overlapping Terrain II	93
5.3. Countries with Muslim Population	94
5.4. Overlapping Terrain III	96
7.1. Sustainability Assessment Map for a Proposed Nuclear Power Station	165
7.2. Sustainability Assessment Map for a Proposed Energy Conservation Scheme	166
7.3. Sustainability Assessment Map Differentiating the Nuclear and Energy Conservation Options	167

Tables

5.1. Citations to Research on Islam, Geography, and Environmental Ethics	96
5.2. Varieties of Muslim Geographic Science	97
5.3. Varieties of Islamic Ethics	99
5.4. Epistles (*Rosa'il*) on Physics	102

Acknowledgments

The present volume was supported by the Eco-Research Chair in Environmental Risk Management at the University of Alberta. The Eco-Research Chair is funded by the Tri-Council Secretariat representing the Medical Research, the Natural Sciences and Engineering Research, and the Social Sciences and Humanities Research Councils of Canada, as well as twenty-four public and private sponsors. We are grateful to Steve Hrudey, holder of this Eco-Research Chair, for his encouragement of our efforts. We would also like to thank the Department of Philosophy at the University of Alberta and the Department of Geography at Texas A & M University for additional support. At Rowman & Littlefield, we thank Julie Kuzneski, our production editor, and especially, Jennifer Ruark, acquisitions editor, who more than anyone else has been an unqualified supporter of our initiative to start this annual.

Most importantly, as is the case with any peer reviewed publication, we are grateful to our referees who reviewed and provided comments on the numerous manuscripts submitted to this first volume. In addition to our associate editors, and many members of our editorial board, we wish to thank the following colleagues for serving as manuscript referees: Allen Carlson, Craig Hanks, Eric Higgs, Joel Kovel, David Macauley, Kelly Parker, Alan Rudy, Benjamin Shippen, and Michael Zimmerman.

Introduction: Geography, Philosophy, and the Environment

Andrew Light and Jonathan M. Smith

We expect people to ask, upon picking up this volume, "Why philosophy and geography?" It is a fair question. The need for a forum, indeed an annual publication devoted to this conjunction, may not be apparent to some readers. We are cautiously confident that for many geographers the combination of the disciplines will not seem too odd, long used as they are to discussions of philosophical issues in the form of critical and methodological questions. But we suspect that many philosophers may dismiss the philosophical relevance of any geographical topic, alien as geography is to the normal territory of philosophical discourse. Some justification, therefore, both of this new peer-reviewed annual publication, and of its inaugural volume, is warranted.

Society for Philosophy and Geography

The story behind *Philosophy and Geography* begins with the formation in 1994 of the Society for Philosophy and Geography (SPG). This organization was started for the single purpose of bringing philosophers and geographers together to exchange views on topics of mutual interest. But again, why start such a society?

The approximately 150 founding members of the society seemed to agree that in some important areas of inquiry, including, for example, issues concerning environmental theory, there had been a marked increase in the interests of philosophers and geographers in each others' work over the past decade. Many environmental ethicists appreciate the synthetic approach of geographers and have found their philosophy improved by the concrete, grounded examples of geographic work. Geographers, for their part, have been eager to move beyond description and explanation of the earth as the home of humankind and undertake critical evaluation of the normative bases of environmental preservation. In general, while there has not yet been much formal discussion directly between the two disciplines, there has been some interesting work at the border of the two fields.[1] It is very common now to hear members of both disciplines discussing issues concerning the status of spatiality, lived space, and theoretical considerations on landscape. Some authors in both fields have become standard reading in any serious work in cultural geography or environmental philosophy.[2] There is ample evidence that some philosophical questions, such as those about the role of cultural and national identity in political theory, cannot be fully answered without consideration of space, place, and landscape.[3] Some schools of geography could not possibly get their research programs off the ground without some reference to philosophical arguments (especially cultural and political geography); some philosophers have come to accept that discussions of space would necessarily always be a component of their work.[4]

The initial success of the society was impressive.[5] It quickly became clear that such a structured union between the two fields would enrich ongoing work in social and political philosophy, philosophy of science, philosophy of technology, iconography of landscape, social construction of nature, and the exploration of cultural geographies, as well as help stimulate new work in other areas. It was also clear that the cluster of ideas that emerges by thinking through topics that would intuitively be of interest to philosophers and geographers also attracted the attention of scholars in other disciplines. The founding members of the society included North Americans and Europeans from a variety of fields beyond the core groups, including sociology, anthropology, political science, public policy, urban and regional planning, architecture, and English and comparative literature. The intent of the founders of the society to avoid anything that looked like an attempt to create a new hybrid discipline, but instead only to encourage more interdisciplinary work on a general range of questions, appeared to be successful.

Philosophy and Geography

Philosophy and Geography was soon proposed to serve as a conduit for the dissemination of new ideas generated by the SPG, and to promote rigorous and innovative new work at the conjunction of these two disciplines that would appeal to a wide audience. Volumes are to be published every year dedicated to the particular topic that forms the general theme for the majority of SPG functions throughout the prior academic year. Submission of manuscripts to the annual is open to anyone, regardless of membership in the SPG or participation in its activities. Each paper is peer reviewed by at least one philosopher and one geographer.

The emphasis of the annual is on the conjunction of the interests of these two disciplines as equals, rather than on the more traditional philosophical predication of another field as an object of philosophical study. We are not trying to create a philosophy *of* geography in the same vein as other philosophers have created subfields of philosophy, such as philosophy of science, philosophy of social science, philosophy of economics, philosophy of biology, and philosophy of physics. Many of these fields often focus exclusively on the philosophical problems of explanation involved in the methodologies of the disciplines under scrutiny. Though this approach has produced some impressive philosophical work, it has not always been of interest to scholars from the fields under study. While, for example, the issue in the philosophy of biology of the status of biology (specifically evolutionary theory) as a falsifiable science under the positivist and postpositivist program of describing the theoretical unity of all science is very interesting, it is easy enough for biologists to go about their work unencumbered by this debate.[6] This is not a failing of philosophy of biology, only a limitation of its audience—a limitation that most philosophers would be willing to accept. Other "philosophy of X" fields fare better when it comes to interdisciplinary activity, for example the successful integration of philosophy of physics as an area of interest, even collaboration, between philosophers and physicists.[7] Not only is philosophical work of interest to physicists in this field, but also the work of physicists is directly relevant to the arguments of philosophers in a way that is not always true for other philosophical studies of other disciplines. We believe that exchanges between philosophers and geographers can yield similarly mutual benefits.

Certainly the difference between some of these fields is explainable by the content of the object of inquiry. Perhaps more important for our present purposes is the evolution of some of these fields from "philosophy of X" to "philosophy *and* X." This is certainly the case for the philosophical study of literature, moving from early philosophy of literature—arguments for how literary texts could serve as a vehicle for philosophical arguments, or the source of examples for philosophical controversy—to philosophy and literature—a forum where texts serve as the repository of philosophical insights themselves, suggesting that their narratives are philosophical arguments in their own right.[8] Such a move is a shift away from the priority of philosophy in the interaction of these two fields. The goal appears to be to promote an openness to the possibility of the X under consideration serving as much as a source of philosophical work itself as it is a field onto which certain antecedent philosophical problems may be scrutinized. In a way, in the new intersection between philosophy and literature, the two meet each other on the same terrain in the spirit of mutual benefit.

Geographers have been no less ready to prioritize, or at least to separate, philosophical issues away from their own work as philosophers have with other disciplines. There has been a distinct lack of effort to articulate geography as a contributor to philosophical debates in its own right. More common has been the geographer's articulation of how some categories of philosophical analysis may be used to discern different disputes in the subdisciplines of geography. This is arguably the case with one of the most important predecessors to our series, Stephen Gale and Gunnar Olsson's collection *Philosophy in Geography,* which was published as volume 20 of Reidel's *Theory and Decision Library,* a series dedicated primarily to philosophical work.[9] All of the contributors and the editors of this volume are geographers. From the opening lines, the editors self-consciously separate out philosophical issues from geographical inquiry:

> Within the scientific community, change is often foreshadowed by an increased concern with philosophical and methodological issues. We find that tradition, principles, and alliances—the very theoretical and epistemological presuppositions of an area of inquiry—come to be questioned and scrutinized. Radical transformations of purpose and substance are proposed. Of equal significance are the moves for modifications in the structure of arguments.[10]

This view of the relationship between philosophy and geography is certainly appropriate for the stated purpose of Gale and Olsson's volume—tracking philosophical shifts in geographical thought—but it is important to note how far removed such a sense of the relationship between the two disciplines is from our initiative. Though we welcome work that does track philosophy in geography, we are not solely interested in promoting that sort of endeavor. Exclusive focus on such a relationship between the fields may result in a too-narrow conception of the contribution of geographers to philosophical problems. Gale and Olsson introduce their justification for listing the papers in their volume in alphabetical order by providing three alternative structures to the collection around three competing philosophical disputes in geography as a field. The disturbing aspect of this arrangement is the suspicion that the editors wouldn't find the intersection of philosophy and geography outside of such permutations legitimate. The ability to describe geography's methodological disputes in philosophical terms appears to be the only thing that legitimizes the conjunction of the two fields. Even though Gale and Olsson offer that the philosophical disputes represented in their volume can be described in a variety of ways—as questions of inference in explanation, as functional "attitudes," and as either "analytic" or "phenomenological"—they still reduce the contributions of the volume to warranting philosophical attention in a very limited sense. Geography and philosophy are not involved here in a mutual conversation; what has been produced is fundamentally a philosophy of geography in the tradition of philosophers' philosophy of X.

We want to avoid such reductive relationships between philosophy and geography in this publication. We have provided here a forum for discussion of issues that will be of interest to members of both disciplines without preference for any arrangement of the possible relation between the two fields. We do not expect that a new field will arise from this conjunction, but only that the work on specific topics of interest to many different scholars will be improved by their interaction in one forum.

Environment as Space, Place, and Idea

Assuming that this account of the origins and orientation of this publication is satisfactory, we turn to the topic at hand, the one covered in this particular volume. It is clear from even a cursory understanding of both disciplines that environmental ethics is a topic of much interest, so it makes sense for the first volume of this series to focus on the environment.[11] But there are also reasons for bringing together the work of philosophers and geographers in this area that follow from the particular issues raised by the demands of environmental studies itself.

Over the last decade, Bryan Norton has made a good case that environmental ethics has suffered from an inability to produce theories that have significantly helped in the creation of better environmental policies.[12] In large part, Norton's case has been structured on the myopic adherence of environmental ethicists to a very narrowly circumscribed philosophical view of what an adequate environmental ethic would look like. But even if one disagrees with Norton's claims about the current lack of utility of philosophy to environmental practice, the implication of the argument that environmental theories ought not be overly construed with a single discipline can still be well taken. Surely Norton's argument could be equally applicable to any form of environmental theory that is grounded solely within the purviews of one discipline. Living up to the rigors of well-established disciplinary demands is certainly commendable. But given the object of environmental theory—the near critical state of the environment and the human relationship to it—it is plausible to suggest that the test of good environmental work would not solely rest in a narrowly prescribed view of a single kind of scholarly merit. It might even be the case that adherence to some strict disciplinary expectations may hamper the creation of productive work in the field.[13] The Norwegian environmental philosopher Sigmund Kvaløy put this point best when he argued that environmental theorists "should strive to be as wide in scope as the attack on the life struggle of the ecosystem and human society is today."[14]

The chapters in this volume contribute to an attempt to meet this challenge by providing interdisciplinary approaches to environmental problems. But rather than trying to write articles that bridge the disciplines of philosophy and geography, authors have been encouraged to write pieces that would be of interest to philosophers, geographers, and others, without sacrificing scholarly merit as determined by their own disciplines. Each contribution is securely grounded within a rigorous tradition. The authors investigate normative questions about environmental issues through different approaches: some theoretical, others empirical, and some combining the two. The result is a unique collection that demonstrates how the study of environmental problems is strengthened by reading, commenting on, and sometimes even taking up each others' work. Though we have certainly not achieved a full integration of philosophy and geography on equal terms with this volume (or with the creation of this annual series), we hope that we have made some kind of contribution to a more comprehensive study of this particular, important, topic.

Overview of the Volume

An effort to bring geographical understanding to bear on philosophical matters is evident in Robert Burch's "On the Ethical Determination of Geography: A Kantian Prolegomenon," in which he interprets the great philosopher's long-standing interest in geography. Geographers have been aware of Kant's contribution to the modern definition of geography as the study of areal integration, but they have not for the most part asked what drew Kant to geography in the first place. Burch makes a compelling argument that Kant appreciated geography as a form of understanding that organized experiences of nature into a coherent whole. Geography did not offer knowledge of nature, but it was nevertheless a way of knowing nature. This geography, understood as the experience of place, was therefore relevant, in Kant's mind, to any philosophy concerned with the place of experience.

The connection between place and philosophy underlies Eric Katz's "Nature's Presence: Reflections on Healing and Domination," which begins with reflections triggered by the author's experience of two places, the Jewish cemetery in Warsaw and the death camp at Majdanek. These reflections lead him to thoughts on healing and forgetting, two forms of restoration. His central questions are whether it is possible for humans to heal the scars of nature, or for nature to obliterate the memory of human evil. Katz concludes that, although this may appear to happen, restoration and atonement are impossible because the processes of both nature and history are irreversible.

The link between political philosophy and the stuff of ordinary, everyday geography is found in Zev Trachtenberg's essay, "The Takings Clause and the Meanings of Land," which examines the various meanings that have been attached to land and their legal ramifications. His central concern is the extent to which government ought to be able to restrict the use of private property without compensating the owner for lost utility. He develops this argument with careful assessment of three coexisting but somewhat contradictory ways in which land is understood: as fungible property, as personal property, and as an object of Aldo Leopold's land ethic.

Many writers contend that traditional human attitudes toward the environment are, at best, irrelevant to the modern environmental crisis, and call for new forms of environmental appreciation, evaluation, and regard. Others profess to find in traditional attitudes the seeds of the present crisis, and consequently mix powerful critiques of the past with their proposals for the future. A third group finds in such arguments yet more evidence of the modern hubris and faith in progress that, in their estimation, fuels environmental destruction. James Wescoat, Jr., draws our attention to such a tradition in "Muslim Contributions to Geography and Environmental Ethics: The Challenges of Comparison and Pluralism." This is particularly interesting as a contribution to the emerging dialogue between philosophy and geography because it reminds us that just as there are philosophies of geography, there are also geographies of philosophy.

In John Clark's, "The Dialectical Social Geography of Elisée Reclus" we find philosophy and geography united in the person of the remarkable French geographer and anarchist. Reclus set out to interpret the place of humans in the natural world and concluded that nature and human nature were locked in a dialectical, mutually determining struggle. Reclus's achievement, in Clark's account, was to offer a description that neither exalted humanity to the role of nature's lord nor debased humanity to the role of nature's slave. This ecophilosophy was ultimately rooted in the development of compassion for humans, and for the various entities of the natural world, and this compassion grew from a fundamentally geographical appreciation of their interdependence.

Much contemporary debate centers on questions of evaluation of nature. In some accounts this is optimally performed by the market, an institution that, however imperfect, is nevertheless the best this sublunary world has to offer. Opposed to this are various arguments for market failure, which is said to have occurred when long-standing demands are not satisfied by supply. Clive Spash and Anthony Clayton argue the second position in "The Maintenance of Natural Capital: Motivations and Methods," which approaches the question of value in nature by way of the somewhat ambiguous metaphor of natural capital. They identify a central cause of misunderstanding between economists and ecologists in their different understanding of sustainability. For economists, the thing to be sustained is output, and this may be accomplished along with environmental degradation and resource depletion through technological innovation and factor substitution. For ecologists, the thing to be sustained is the productive system itself. The authors conclude that we cannot rely on the market or scientific experts to evaluate nature because both institutions are blind to certain values in nature. Notable among these is the value of reduced future risk that is created by present-day caution, and humble estimation of our ability to fully comprehend or efficiently manage complex ecosystems.

Roger Paden takes up some similar themes in "Wilderness Management." After reviewing the three dominant forms of wilderness policy—exploitation, conservation, and preservation—Paden undertakes a critical assessment of preservation and some of its latent weaknesses. Preservation, or minimal management of wilderness, is justified on the basis of imperfect human understanding, and a consequent inability to predict the full outcome of any particular intervention in nature. This justification will be weakened, if never perhaps fully eliminated, as scientific knowledge improves. This is not a problem for biocentric arguments, which view preservation as a means to protect the interest that nonhumans have in their own autonomous and spontaneous development. The biocentric argument for wilderness preservation faces its own problems, however, as a result of the shift in emphasis from preservation of natural products (endangered species), which can be construed as having interests, to the preservation of natural processes (wildfires), which cannot.

Every essay in this volume is confronted with the problem of deep ambiguities in the meaning of environment, a concept of which there are several varieties. In "Mead and Heidegger: Exploring the Ethics and Theory of Space, Place, and the Environment" Eliza Steelwater claims that these meanings have always been subject to negotiation. In an unconventional and provocative argument, she draws on the work of Martin Heidegger and George Herbert Mead and insists that any environmental ethic must be grounded in purposeful relations with the phenomenological world of experience. The origin of a proper regard for the environment is to be found within the horizon of personal involvement, care, and concern.

Roger King arrives at a similar conclusion in "Critical Reflections on Biocentric Environmental Ethics." He opens his argument with a criticism of biocentrism, which is flawed, he claims, both practically and epistemologically. As a critique of human domination of nature, biocentrism has little to say about treatment of that large part of nature that has been, and will be, put to work in the service of human needs. Furthermore, its epistemological presumption of disinterestedness is, like all such claims to impartiality, highly suspicious. Instead, King proposes an ecofeminist solution that is grounded in the position and point of view of particular moral beings. Environmental ethics should not be derived from universal principles, but rather should emerge in an ad hoc manner as particular issues arise in experience.

Matthew Gandy's "Ecology, Modernity, and the Intellectual Legacy of the Frankfurt School" stands somewhat apart from several of the others in this volume because he is reluctant to blur the boundary between humans and the natural world. Drawing on the tradition of German critical theory, and particularly contemporary writers like Jürgen Habermas and Ulrich Beck, Gandy argues for an ethics grounded in communicative rationality. Our present problems will not be solved by reforming the relations of humanity and the natural world, he argues, particularly if responsibility for these reforms is placed in the hands of experts. Indeed, it is the rise to power of these technocrats, and the consequent withering of practical reason, that has brought us to the point where, as Beck has argued, science is the source of most new environmental problems as well as solutions.

We take it as an encouraging sign of the vitality of environmental ethics that all of these essays have offered criticisms of some of the field's central assumptions. Iconoclasm outweighs dogma by a large margin. Nowhere is this more evident than in Annie Booth's "Critical Questions in Environmental Philosophy," which is a spirited critique of bioregionalism, ecofeminism, and the veneration of indigenous peoples. Booth questions the relevance of bioregionalism to a world that supports a vast human population only through productivity gains achieved by regional interdependence. She rejects ecofeminism as a position founded on accusation and condemnation, divisive tendencies that begin with renunciation of solidarity with the male half of the species. Finally she criticizes those who seek guidance from indigenous peoples and their supposed reverence for nature, since this tutelage depends on a myth that obscures the present plight and aspirations of those same native peoples.

These eleven articles were selected via a rigorous peer review process. Nevertheless, the volume is integrated by several persistent themes, and two of these are of particular importance. The first is clearly the concept of a moral community and the boundary that divides it from those things understood as having only instrumental value. This is almost invariably described with spatial metaphors of circles and centers, closeness and distance, and, of course, boundaries. The second is the concept of place, since each one of us is environed in a particular place, and it is in this place that we act upon, suffer, and understand our environment. Space and place are, of course, at the heart of geography, but they are not the sole property of that field; indeed, it appears clear to us that they are concepts equally indispensable to the geographical and the ethical imaginations.

Notes

1. Just a few examples of the more philosophical work of geographers include: Peter Jackson and Jan Penrose, eds., *Constructions of Race, Place, and Nation* (Minneapolis: University of Minnesota Press, 1993); James Duncan and David Ley, eds., *Place/Culture/Representation* (London: Routledge, 1993); Anne Buttimer, *Geography and the Human Spirit* (Baltimore: The Johns Hopkins University Press, 1993); Trevor J. Barnes and James Duncan, eds., *Writing Worlds: Discourse, Text and Metaphor in the Representation of Landscape* (London: Routledge, 1992); J. Nicholas Entrokin, *The Betweenness of Place: Towards a Geography of Modernity* (Baltimore: The Johns Hopkins University Press, 1991); Edward Soja, *Postmodern Geographies: The Reassertion of Space in Critical Social Theory* (London: Routledge, 1989); David Harvey, *The Condition of Postmodernity* (London: Basil Blackwell, 1989). Certainly the work of Denis Cosgrove, Neil Smith, David Smith, Yi Fu Tuan, John Pickles, and others could be mentioned as well. There are relatively fewer philosophers who have worked directly on geographical themes, though one could include Henri Lefebvre, Albert Borgmann, Edward Casey, John McDermott, Larry Hayworth, David Abram, and to a limited extent John Sallis, in a list of those writing directly on these themes. Most environmental philosophers and philosophers of technology, however, have approached geographical issues in their work, though often without explicit reference to the literature in geography. The one certain exception is Bryan Norton, who has begun writing collaborative papers with geographers.

2. For those working on ideas of wilderness, for example, no study from either discipline would be complete without a close look at the philosopher Max Oelshlaeger or the geographer John Rennie-Short.

3. See for example the final chapter of Iris Marion Young's groundbreaking book on identity politics, *Justice and the Politics of Difference* (Princeton: Princeton University Press, 1990). Young attempts a description of the just city as an expression of a complete politics of difference. Certainly such attempts have roots in the geography literature including David Harvey's singular achievement in *Social Justice and the City* (Baltimore: The Johns Hopkins University Press, 1973). For more recent work that goes well beyond class as a component of spatial identity, see Michael Keith and Steve Pile, eds., *Space and the Politics of Identity* (London: Routledge Press, 1993).

4. This seems very clear for Habermasians who would no more ignore the importance of public space in a complete description of a democratic society than they would the role of law. See Jürgen Habermas, *The Structural Transformation of the Public Sphere* (Oxford: Polity Press, 1989).

5. In only the first year of its existence, the society scheduled ten paper sessions, which met in conjunction with all three divisional meetings of the American Philosophical Association, the Association of American Geographers, and the Canadian Learned Societies meetings. Even more sessions were held in the second year.

6. See, for example, the ways these issues are raised in Alexander Rosenberg, *The Structure of Biological Science* (Cambridge: Cambridge University Press, 1985), and more recently the evolution of the debate on the unity of science in Rosenberg's *Instrumental Biology, or, The Disunity of Science* (Chicago: University of Chicago Press, 1994).

7. See Steven Savitt, ed., *Time's Arrow Today: Recent Physical and Philosophical Work on the Direction of Time* (Cambridge: Cambridge University Press, 1995).

8. One could make the case that Martha Nussbaum has the most refined form of philosophy of literature in her earlier *The Fragility of Goodness: Luck and Ethics in Greek Tragedy and Philosophy* (Cambridge: Cambridge University Press, 1986), and the more recent, *Love's Knowledge: Essays on Philosophy and Literature* (Oxford: Oxford University Press, 1990). Richard Rorty has taken philosophy and literature to an extreme even to the point of suggesting that philosophy is simply a form of literature, or perhaps more accurately, that literature is a better repository of philosophical claims than philosophy. See Rorty, *Contingency, Irony, and Solidarity* (Cambridge: Cambridge University Press, 1989). Also see his "Truth and Freedom: A Reply to Thomas McCarthy," *Critical Inquiry* 16, no. 3 (1990): 633–43.

9. Stephen Gale and Gunnar Olsson, eds., *Philosophy in Geography* (Dordrecht: D. Reidel Publishers, 1979). The volume includes some of the previously noted geographers who have dealt with philosophical themes in their work, such as David Harvey and Yi Fu Tuan.

10. Gale and Olsson, *Philosophy,* ix.

11. We will not attempt here a defense of the claim that both philosophers and geographers are interested in environmental issues, beyond a few anecdotal suggestions. In philosophy, consider, for example, that almost every major textbook publisher in North America has published an introductory textbook on environmental philosophy. Additionally, almost every major university publisher has produced a major scholarly work in the field. There are also a wealth of special collections on specific topics such as ecofeminism. There are at least five major journals in the field (including *Environmental Ethics,* which will soon be twenty years old), and a host of other journals have produced special issues on environmental philosophy or political theory. In geography the centrality of the environment hardly needs justification, though we think it noteworthy that the journal *Ecumene* was started by two geographers in 1994 to publish scholarly research on various aspects of the cultural appropriation of nature, landscape, and the environment in an explicitly normative context.

12. See Bryan Norton, *Why Preserve Natural Variety?* (Princeton: Princeton University Press, 1987) and *Toward Unity among Environmentalists* (Oxford: Oxford University Press, 1991). See most recently Norton's "Integration or Reduction: Two Approaches to Environmental Values," in *Environmental Pragmatism,* ed. Andrew Light and Eric Katz (London: Routledge Press, 1996), 105–38.

13. Even with the aforementioned dramatic growth in environmental studies, philosophers at least should not fool themselves into believing that environmental ethics is anywhere close to the center of their discipline. Many mainstream philosophers will still maintain to this day that environmental ethics is not a serious, or even interesting, form of philosophy. Again, we will not move beyond anecdotal evidence for this suggestion. Consider the lack of full-time academic positions in environmental ethics in relation to the number of actual classes taught on the subject, as well as the hostility that many environmental philosophers perceived to be the standard philosophical climate toward environmental issues. For a very early account of this last tension see Richard Sylvan, "Do We Need a New, an Environmental Ethic?" reprinted in *Environmental Philosophy*, ed. Michael Zimmerman et al. (Englewood Cliffs, N.J.: Prentice Hall, 1993), 12–21. The point is that if environmental philosophers strictly adhered to some interpretations of the requirements of their broader discipline, they could find reasons to give up on doing environmental ethics in the first place.

14. Sigmund Kvaløy, "Complexity and Time: Breaking the Pyramid's Reign," in *Wisdom in the Open Air: The Norwegian Roots of Deep Ecology*, ed. Peter Reed and David Rothenberg (Minneapolis: University of Minnesota Press, 1993), 119.

On the Ethical Determination of Geography: A Kantian Prolegomenon

Robert Burch

> Kant is less a prisoner of the categories of subject and object than he is believed to be, since his idea of the Copernican revolution puts thought into a direct relationship with the earth.
> —Gilles Deleuze and Felix Guattari

I

The immediate topic of the following discussion is the relationship between philosophy and geography and the question of what ethical significance that relationship might have. To help situate this topic in the current intellectual landscape, and to indicate what particular direction the discussion will take, some orientation is needed.

If we approached the topic in a time-honored way, we might begin by formulating some essential definitions of philosophy and geography, using these in turn to set the two disciplines into a definite relationship in order specifically to see what ethical precepts might thereby be adduced. The method of this approach would itself be philosophical in a broad sense, since it would be philosophy, and not geography, that would decide upon the essential definitions in both cases, establish the relationship between the disciplines, and adduce the relevant ethical precepts. In this process, then, geography would have no independent voice.

What worth this appraoch may have, it now seems all but closed.

Philosophy no longer enjoys unchallenged authority as the sort of legislator and sheriff that lays down the law for all possible sciences and polices their respective methodologies and borders. Such a rigidly authoritarian conception of philosophy's critical role has in recent decades largely given way to talk of "paradigms" and competing local rationalities. Moreover, on the wider intellectual scene, it has lately become fashionable to deny philosophy such critical authority even over itself. Traditionally, philosophy had claimed exclusive privilege as the only truly self-defining and self-legitimating discourse, a homologous quest for an all-comprehensive "truth" pursued as an end in itself. Contrarily, the height of current fashion is to disclaim the very ideal of a thoroughly self-possessed discourse, systematic comprehension, or unadulterated will to Truth. Although initially exotic, the variety of textual and "genealogical" strategies used to effect this denial are now widely familiar. What they have in common are their disparate and "polyvocal" attempts to expose everywhere heterologous and thus self-subverting dimensions in the very construction of philosophical discourses, texts, and theories. Where such strategies hold sway, philosophy is left with nothing particularly authoritative to say, such that in its traditional conception and practice it is displaced as a discipline.

In the case of geography, displacement of the discipline seems less a current threat than a persistent predicament. This has to do in part with geography's eclectic sources. "Geographic ideas . . . almost invariably are derived from broader inquiries like the origin and nature of life, the nature of man, the physical and biological characteristics of the earth. Of necessity they are spread widely over many areas of thought."[1] Fully appreciative of this aspect of geography's provenance, a recent scholar goes a step further. He characterizes geography as inherently a "contested enterprise" and draws from its history the skeptical lesson that there is no "essential nature of geography . . . eternally fixed," but only "a situated geography" whose " 'nature' . . . is always negotiated."[2] Now, to concede to such indeterminacy might well serve geographers' strategic interest. With their discipline freed from having to conform a priori to some concept of a fixed essence imposed *ab extra* by philosophical fiat, geographers might be able to negotiate the "nature" of geography freely for themselves *in medias res*, speaking directly in the name of a distinctive geographical perspective and truth. But then again, such indeterminacy might serve them poorly. For to have an effective voice in any such negotiations still requires some nonnegotiable reasons already in place to listen seriously to the voices of those who would speak for any particular perspective and truth or any particular configuration of knowledge. As the current restructuring of university faculties attests, merely believing in yourself and in your own theoretical approach holds scant assurance of being heeded to one's advantage. *Plus ça change, plus c'est la même chose.*

Yet even if in some rather old-fashioned way essential definitions and philosophy's authority were both firmly in place, it would still not be clear how a turn to geography per se *could* yield any distinctive ethical precepts. For one's general ethical stance would still be worked out in advance on philosophical grounds. Depending on the sort of stance taken, one might then look to the established results of geographical research as a locus of ethical concerns or a source of ethically relevant information. But the stance itself and information provided would be externally related, and the relationship would be strictly asymmetrical. To argue from the empirical results of geographical research directly to ethical injunctions and actions would be to commit a textbook fallacy. In general, where one's ethical stance is to be more than ideological posturing, it must be supported by philosophical argument that transcends "empirical facts" as such.

Of course, one might dismiss our topic altogether by embracing wholeheartedly the postmodern fashion. Postmodern discourses disavow any sort of singular "logic" that would accord ultimate authority to one way of thinking or that would provide "maps" of topics that are to correspond to what is essential. In this sense, postmodern thinking is not so much a definite "position" opposed to philosophy and modernism, as an anarchical "dis-position" that operates subversively in and around such discourses at different levels and along different trajectories. "Simplified in the extreme, we take for 'postmodern' incredulity with regard to *métarécits.*"[3] Yet herein lies the rub. Such thinking cannot go so far as to be incredulous about its own incredulity, nor then about the ethical whither and wherefore of its own critical discourse. Or rather it can go that far, but at the risk of making all discourses, including its own, so utterly problematic as to be a matter of critical indifference, with the effect that nothing in particular truly calls for thinking. Outwardly hypercritical, postmodern discourses remain inwardly hypocritical, unless they can subject their own "dis-positions" to critique and offer some argument, however inconclusive, that explores and defends their own presuppositions, rules, and effects. Yet merely to accept this requirement is to reinstate the authority of a transcending philosophical reason, however much one might will to disbelieve it in all other instances. And to meet this requirement elicits the awareness that all one's critical talk, however fluid and seemingly noncommittal, has its ontological and ethical presuppositions.

The following discussion explores the possibility of an original, that is, ontological, relationship between philosophy and geography by considering philosophically how geography could have a fundamental ontological significance. This in turn will serve to indicate how the relationship could have an intrinsic ethical significance. Although this task is described rather generally, the discussion itself follows a particular historical path, seeking to "retrieve" the problem of the relationship of philosophy and geography specifically from Kantian thought. The idea of a "retrieval" (*Wiederholung*) is borrowed here from Heidegger. "By the retrieval of a fundamental problem," Heidegger writes, "we understand the disclosure of its original, hitherto concealed possibilities, through the working out of which the problem is transformed, and so first comes to be preserved in its problematic content."[4]

At one level, it is Kant's acknowledged historical position that justifies this recourse to his thought. That Kant's Copernican revolution redirected the very course of modern philosophy needs hardly to be argued. Less well-known is the fact that Kant has had a considerable influence too on the development of the discipline of geography. As recently as 1968, geographer Preston James wrote, "Kant . . . gave geography its place in the over-all framework of organized, objective knowledge (sciences)," and it "is essentially [his] concept of the place of geography among the sciences that has guided the main stream of geographic thought since Kant." Others have made this point even more forcefully. "Kant's contribution . . . consisted of his definition of the nature of geography and its relationship to the natural sciences. . . . Confusion about the aim and content of geography has almost always only appeared when Kant's analysis has been ignored." And still others have claimed that it is Kant's "philosophical construction" that served to establish for geography "an honorable status among the sciences," and from which "the fundamental justification for geography" has been derived.[5]

At a deeper level, however, there is a philosophical reason for turning to Kant. It is the conviction—which only the discussion itself can bear out—that there is a fundamental tension in Kant's conception of the relationship of philosophy and geography, and that the resolution of this tension opens the way to a radical "existential" reconfiguration of how philosophy, geography, ethics, and ontology interrelate.

II

In the curricula of the modern university, the philosophers' interest in geography antedates the firm establishment of the discipline itself by some one hundred years.[6] In this regard, it is Kant who is the innovator. In the summer of 1756, in his second semester of university teaching at Königsberg, Kant adds to his lectures on logic, mathematics, and metaphysics, a course on physical geography.[7] Now, however routine such a course may be in today's university, to Kant's colleagues and students it would have been quite novel. Although the sixteenth and seventeenth centuries had witnessed in Europe a revival of interest in theoretical geography (evidenced, e.g., by Mercator's maps and Apianus's *Cosmographia*, and later by Varenius's *Geographia generalis*, Clüver's *Introductionis in Universam geographim*, and Carpenter's *Geographie Delineated Forth in Two Bookes*), it was only by the mid–eighteenth century that official courses on geographic topics were beginning to be introduced in German universities; and "for Königsberg University in 1756, geography was something entirely new."[8]

Kant was not the first modern philosopher to pay some attention to geography. There were philosophers before him who had included geographic knowledge as part of general enlightenment and worldliness; others who had taken account of geography in the course of a general reconfiguration and reconstruction of the sciences; and still others who had regarded geographical considerations as important in a more general theorizing.[9] But no philosopher before Kant had explicitly set out to determine philosophically the nature and role of geography as such, nor to establish specifically the place of the discipline with respect to the "origin" and "arrangement" of "the whole of our knowledge."[10]

It would seem, moreover, that Kant's interest in geography was more than a phase and a dalliance. During his forty-year teaching career, he would repeat his course on the subject no less than forty-eight times, continuing to give it regularly, even after his appointment to professor (1770), and even after, so to speak, he truly becomes "Kant" (1781). Only his lectures on logic (fifty-four times) and metaphysics (forty-nine times) were offered more frequently, with the lectures on ethics (twenty-eight times) counting a distant fourth.[11] Yet there was no official charge to Kant to give this particular course. Rather he had to flout the official policy that required one to lecture by commentary on a state or university approved text, and lecture instead on his "own essays" and his "own notes."[12] In later years, Kant would even require a special state exemption to continue this practice in this course.[13]

Now, one can imagine that the youthful Kant, financially strapped and suffering from the "hard labor" of a lecturing regimen that by his own admission effectively stifled any "nobler impulse," might well have decided that offering a course on physical geography made good sense.[14] It was meant, after all, to be a "popular lecture" that would attract "people of other professions." Moreover, the subject was one on which Kant had already published, in which throughout his life he maintained an abiding avocational interest, and which he thought "provided entertainment and rich material for social conversation." But that over the course of his career the author of *The Critique of Pure Reason* should have lectured on geography virtually the same number of times as on metaphysics, and that the author of *The Foundations of the Metaphysics of Morals*, *The Critique of Practical Reason*, and *The Metaphysics of Morals* should have lectured on geography *twenty* more times than on moral philosophy, is surely passing strange. This sense of strangeness is strengthened, if one considers that Kant was awakened from his "dogmatic slumber" and embarked on a "new direction" in philosophy rather late in his career, and that he came increasing to feel that he was writing against death. That he should then have given such time to lateral concerns, however worthwhile in themselves, is difficult to imagine. If, then, we take Kant seriously as a philosopher, we are obliged to look for a convincing *philosophical* rationale for his persistent attention to geography.[15]

III

Yet to find a convincing philosophical rationale in the received, textbook account of Kant's system is no easy task. Kant, we recall, defines philosophy formally, scholastically, as "the system of . . . rational cognitions from concepts."[16] Rational cognitions, he claims, are in general those cognitions "obtained *ex principiis*,—out of certain principles," being specifically either mathematical (i.e., from the "construction of concepts," where the object of knowledge is "exhibited a priori in intuition"), or philosophic (i.e., "from concepts" alone).[17] To rational cognition, thus characterized, Kant juxtaposes "historical" knowledge, by which he means cognitions "obtained *ex datis*,—merely from experience." Taken together, the "rational" and the "historical" comprise exhaustively the "two modes" into which, "according to their *subjective* origins," human knowledge divides, just as, "according to their objective origins . . . all cognitions are either rational or empirical."

It is not necessary to consider this schema in detail to see that strictly in terms of it, geographic knowledge, whether according to its subjective or its objective origins, would have no *intrinsic* philosophical significance. One's geographic circumstances may favor or hamper one's going about the work of philosophizing, but in either case they are accidental to the philosophical truths one seeks. As systematic rational knowledge from concepts, philosophy transcends all empirical knowledge *ex datis* and every empirical question of *quid facti*. Geography and philosophy would thus have nothing *essentially* to do with each other.

This points to a second sort of difficulty. Like Carpenter before him, Kant defines (physical) geography according to the "ordinary meaning" of the term as "description of the earth" (i.e., literally, *geographein*).[18] By means of such description, Kant explains, "in general we examine the stage [*Shauplatz*] of nature, the earth itself, and the regions where things are actually to be found."[19] The task is to provide a "physical" classification of things according to their actual location in space, rather than a "logical" classification of things by species and kind. Yet, like Aristotle, Kant too rejects the idea of a *science* of the merely accidental and contingent. Although a description that merely showed where things happen to be might add to the store of "common knowledge," it would not be a "science."[20] A science by Kant's criterion must be a "system," that is, a "unity of a manifold of cognitions under an idea," where an idea is "the rational concept of the form of a totality."[21] If, then, there is truly a *science* of geography, the "idea" or "architectonic" that "creates the science" must be shown.[22]

As *scientific* description of the earth, physical geography is "a general compendium [*Abriß*] of nature."[23] Its founding idea is that of nature in the sense of a world. "The world," Kant explains, "is the substratum and stage upon which the play of our skill [*Geschicklichkeit*] proceeds. It is the ground upon which our knowledge is acquired and applied."[24] Understood in this way, physical geography outlines the pragmatic situating context for all our natural sciences, and tells us where things are to be found, not as facts to be iterated, but as objects of the actual, integral *world* of our experience. As "a description of nature," then, physical geography describes "the *whole* earth."[25]

Judged on these terms, however, the place of geography is not so straightforward. Kant divides the doctrine of nature into natural science "proper" (*eigentlich*), and what he calls the "historical doctrine of nature."[26] The latter, he says, "contains nothing other than the systematically ordered facts of natural things." He in turn subdivides the historical doctrine of nature into "the description of nature as a system of classes of natural things according to their similarities," and "history of nature as the systematic presentation of natural things in different times and different locations." Now, since the former description seems to involve just the sort of classification that Kant claims geography does *not* give, it would not fit there. Yet neither does it quite fit under the other arm. For the history of nature "describes the occurrences of the whole of nature as they have been through all time," whereas geography "concerns appearances that, with respect to space, occur at the same time . . . side by side."[27]

On closer inspection, this exclusion of geography is only apparent. Kant's thesis is that the history of nature proper is a diachronic account of nature made up of a suite of synchronic accounts of nature of the sort that a geographical description would provide. Thus, he claims that "genuine history is nothing other than a continuous geography." There is of course a difference. Geography proper is a "description" of the whole earth as it presently is, albeit not from one experience of the earth as a single object of outer sense, but from "perceptions which taken together, would constitute the experience."[28] By contrast, the history of nature proper is a "narrative." It not only must construct the synchronic moments that make up the history itself, as if each moment were the result of a geographic description, but also to tell the story of "what kinds of changes have been gone through in all periods."[29] Thus characterized, geography and history are said by Kant to "fill up the entire range of our knowledge; geography namely that of space, but history that of time."[30]

That geography holds such a place in Kant's system outside the domain of natural science proper does not seem to enhance its importance. For what characterizes the latter domain is precisely its philosophical foundation. "Natural science properly so called presupposes metaphysics of nature."[31] It does so, since such science treats of the existence of natural things according to their essence, that is, according to "the internal principle of everything that belongs to the possibility of a thing," where "possibility" has to do with the a priori conditions of the thing being an object of experience.[32] Hence natural science proper is a rational doctrine of nature, as opposed to a historical one, and its principles hold of the things of nature necessarily. "Only that whose certainty is apodeictic can be called science proper; cognition that can contain merely empirical certainty is only improperly called science."[33] The latter may explain given facts in terms of laws, but those laws are merely empirical and "carry with themselves no consciousness of their necessity." Such science, Kant says (using chemistry as his example), "should be called systematic art [*Kunst*] rather than science." In Kant's taxonomy, then, geography has a place within that branch of the doctrine of nature that seems the least philosophical, being neither natural science proper nor improper. However, in the body of his lectures on physical geography, Kant does offer natural scientific explanations of the improper kind. He discusses, for example, the "causes" (*Ursachen*) of such phenomena as salt water, kinds of salt, waves, breakers, ocean currents, whirlpools, tides, sand banks, and earthquakes.[34] But again such a function for geography is seemingly quite removed from anything of philosophical significance, either directly or indirectly.

Difficulties concerning the place of geography in Kant's theoretical philosophy can be multiplied further. Formally, Kant defines nature itself as "the existence of things insofar as that existence is determined according to universal laws."[35] Moreover, he considers this view of nature to be "physiological" in the literal sense, that is, as having to do with the fundamental principles (*logoi*) in terms of which we know nature (*physis*) as an object of experience in general. From this physiological perspective, nature *is* the object of experience, and experience *is* empirical knowledge, understood as the knowledge that natural science works out systematically.[36] In *all* investigation of nature from this physiological perspective, the "idea of nature's purposiveness" and its determination according to final causes (either in the internal workings of objects as such, or in their external arrangements) "is neither a concept of nature nor a concept of freedom."[37] In other words, the principles neither of theoretical nor of moral science allow us to claim *knowledge* of nature per se as anything other than knowledge of objects operating under a mechanistic causality. The idea of the purposiveness of nature is merely "a subjective principle (maxim) of judgment," that is, a regulative principle that accords with the intrinsic need of our reason and governs how we proceed in investigating nature, but which "attributes nothing whatsoever to the object (nature)." If, then, our natural, scientific investigations should seem to discover some purposive, systematic unity in nature, this is merely "a lucky chance favoring our aim." In this view, Newtonian physics is the paradigmatic natural science, for which nature is simply matter in motion under universal laws, it being in terms of such laws that all genuine explanation is given.

Judged on these terms, geography clearly does not constitute a natural science in the specific sense of a system of objective knowledge of the "earth" from a "physiological" perspective. Indeed, it is not essentially a *theory* of nature in Kant's sense, since it does not simply "explain natural phenomena in mechanical terms through their efficient causes."[38] Instead geography is a *description* of nature, and as such has for its "aim" a "thoroughly coherent experience." Such description "posits purposes of nature in natural products insofar as these form a system in terms of teleological concepts," but it does not constitute a teleological *doctrine* of nature such that in virtue of it nature is *known* objectively to be truly made up of such purposes. Thus, however valuable such description may be, it "gives no information whatever about the origin and inner possibility of these forms, whereas that is just what theoretical natural science is properly to do."

Likewise, in terms of its subject matter, geography is not a natural science. The "earth," which geography describes, is not (to borrow Husserl's pejorative phrase) a "substructed world of idealities."[39] In other words, it is not a determinate domain of objects that, methodologically, geography sets up a priori in order, then a posteriori to investigate with true exactness and certainty. It is rather the earth as made up of mundane things—trees, rivers, mountains, plains, flora and fauna, peoples—the natural things of our actual immediate environment, part of the horizon of our everyday life and experience. But for Kant the mark of all genuine science is that it "has insight only into that which it produces [*hervorbringt*] after a plan of its own," and that in seeking knowledge of nature, empirical science "must adopt as its guide . . . that which it has itself put into nature."[40] Accordingly, all positive science is in principle scientific precisely in proportion to which, methodologically, it determines the essence of its particular domain of objects a priori. But insofar as geography understands the essence of its subject matter in terms of everyday experience, then qua science it is (to use Kant's own metaphor) "kept in nature's leadstrings" and in that measure is unscientific. Thus, even were geography to provide a "purposeful arrangement of our knowledge," it could do so only in terms that lie outside natural science and to an end that is not strictly scientific.[41]

One encounters similar ambiguities when one considers the relation of geography to Kant's moral philosophy. As every student of philosophy knows, Kant sets up an exhaustive division between the realms of nature and morality. In moral philosophy, moreover, Kant is a strict deontologist. Moral worth consists in acting in accord with the moral law for its own sake. The moral law is an apodeictic principle of universal reason (i.e., the principle of duty), which we, as finite rational beings, apprehend as an absolute command (i.e., the categorical imperative). In acting in accord with the moral law, we act freely, by determining ourselves unconditionally in accord with universal reason. Free, moral action is at any time within a rational being's power; therefore such a being is at all times morally responsible. Morality, then, is a matter of strict universality. The moral law "applies to all rational beings generally" and as such it "must be valid with absolute necessity, and not merely under contingent conditions and with exceptions."[42] It is also a matter of strict autonomy. "The will is not only subject to the [moral] law but is subject to it in such a way that it must be regarded as self-legislative, and only for this reason as being subject to the law."[43] Morality is thus also anthropocentric, or rather more accurately, *Vernunftwesenhaft*-centric. The moral law is formed from the conception of that which is good without exception (i.e., a good will), and hence that which is necessarily an end for everyone. In virtue of being capable of rational self-determination and obligated to actualize this capacity, it is humanity, or rather rational being in general, that is such an end. "Suppose there were something," Kant writes, "whose existence has in itself an absolute value, something which as an end in itself could be the ground of determinate laws, then in it, *and in it alone*, would there be the ground of a possible categorical imperative. . . . Humanity, and in general every rational being, exists as an end in itself."[44]

Albeit brief and superficial, this review of Kant's moral doctrine suggests no obvious place for geography within that domain. The mutual dependence between moral obligation and moral freedom, as well as the very nature of moral worth itself, transcend all contingent, human considerations of place and time, and hence all human geographic conditions and cognitions.[45] "All moral concepts have their seat and origin entirely a priori in reason, and . . . in this purity of their origin lies their very worthiness to serve us as supreme practical principles, and everything empirical added to them is just so much taken away from their genuine influence and from the unqualified worth of [their corresponding] actions."[46] What constitutes good life-conduct remains everywhere and always the same for all rational beings, precisely because the moral law holds absolutely, and moral worth consists in willing the moral law for its own sake. Thus morality is conditioned by no determinate, human spatial/temporal horizon, and no concrete contingent context of human, all-too-human, affairs. Hans Jonas has put this point (disapprovingly) in the following way: "Kant's categorical imperative was addressed to the individual, and its criterion was instantaneous. . . . The principle was not one of objective responsibility but of the subjective quality of my self-determination."[47] On this account, geographic knowledge may serve the *bonum pragmaticum* as a specifically human good, providing a "technical" knowledge of the world that humans can put to use to get what they want and to avoid harm. But it is not knowledge relevant essentially to the *summum bonum*, that is, to the highest good as the necessary end for rational beings in general. And it is in this highest *rational* good, and not in a specifically human good alone, that true moral worth lies.

IV

If, then, there is no obvious philosophical rationale for Kant's persistent attention to geography in the received textbook account of his system, we need to take a different tack, looking perhaps not so much at the explicit structure of the system as at its underlying sense.

Despite Kant's scholastic concern with science and metaphysics, his guiding interest is at bottom ethical. As is well known, in the pursuit of the former, he "found it necessary to deny knowledge in order to obtain a place for [moral] belief," and thereby to establish in the latter sphere the "possibility for the practical extension of pure reason."[48] In a famous marginalia, Kant confesses: "I am myself by inclination an inquirer. . . . [Yet] I should regard myself as far more dispensable than a common laborer, did I not believe that this attitude [i.e., of inquirer] is able to give to all a worth that establishes the rights of humanity."[49] Likewise, invoking the authority of Francis Bacon as an epigraph to the First Critique, Kant confirms that "the real work . . . is building the foundations of human utility and power."[50] Although in the theoretical realm this has to do with building the foundations of the positive knowledge by which we effectively command nature, the paramount task concerns the foundations of morality, in terms of which we command ourselves, realize our peculiar human dignity and worth, and work toward the highest good. If in Kantian terms geography has a genuine philosophical significance, it might be found here.

What needs to be accounted for philosophically is the fact that Kant continually *lectured* on physical geography. It might help then to note that Kant understood his work as a university lecturer differently than his work as the public author of philosophic texts. In the latter case, both the style of presentation (whether technical or popular) and the mode of argumentation were determined by the rational nature of the subject matter itself and the demands of reason alone, the published texts being presented freely before the reading public for scholarly debate and evaluation.[51] By contrast, Kant's academic lectures were for the (private) instruction of students. Thus, they were guided by a pedagogic as well as a strictly "scientific" purpose, and their presentation was subject to the duties and responsibilities that bound a university teacher as one holding a civil post. Thus, even though in the lectures Kant might be critical of this or that opinion, his scholastic purpose is not publicly to advance knowledge or to express his particular point of view, but to offer general instruction in the form of compendia, fleshed out with illustrations and examples, on the received knowledge in the field. In this process, the pedagogic strategy was to "form [*bilde*] first the man of *understanding*, then of *reason*, and finally of *learning*."[52] To this end, then, Kant's lectures typically do not begin with a direct and full exposition of the highest truths in a field, but with a critical clarification of the common rational understanding (and misunderstanding) already operative in experience and the world. (Even the study of logic Kant approaches first as a "critique and preface to sound understanding," a kind of "quarantine the student must observe who wants to pass from the land of prejudice and error to the domain of enlightened reason and science," and only then as "a critique and preface to authentic learnedness."[53])

Yet, far from diminishing the value of the lectures, this approach underscores their real worth. For the final end of the lectures is not just to survey discrete bodies of received knowledge in a scholastic way as knowledge about various sorts of things in the world, but also beyond this instruction to cultivate the practical ideal of a comprehensive human wisdom. The "ideal teacher," Kant claimed, is not an "artificer in the field of reason" who regards philosophy according to "its scholastic concept" as "a system of knowledge . . . sought solely in its character as a science," but one who pursues philosophy according to its worldly concept (*"conceptus cosmicus"*) as "relating to that *in which everyone necessarily has an interest,*" and who, in the ideal, instructs and sets tasks in order "to further the essential ends of human reason."[54] Admittedly, between the starting point and the final goal of this pedagogic process, Kant recognizes that there is no royal road. For the full perfection of the idea of wisdom (i.e., "the property of a will to accord with the highest good as the ultimate end of all things"[55]) exists in finite beings only as an "infinite task," and whereas one can "enrich the natural understanding with concepts and equip it with rules," with respect to the power of judgment, which too is required for wisdom, "it cannot be *instructed*, but only *exercised*."[56] Accordingly, "the appropriate method of instruction in worldly wisdom is *zetetic*, as the ancients unanimously called it (by *zētein*), that is to say, *searching*, and it becomes dogmatic (i.e., preemptory) only with reason already trained in distinct parts."[57]

In the published announcement for his lectures on physical geography, Kant seems to suggest something like this "zetetic" method. After outlining the main divisions of the subject matter to be treated, he remarks: "All this, however, is not with that completeness and philosophical exactness in each part that is a matter for physics and natural history, but with the rational curiosity of a traveler, who everywhere seeks out what is noteworthy, peculiar and beautiful, collates his collection of observations, and reflects on its design."[58] Since the subject is physical geography, the metaphor of the traveler seems apt. Yet it is used in a somewhat narrower sense than a present-day reader might expect.

Although Kant's rationally curious traveler is no physicist or natural historian, neither is he or she the ubiquitous latter-day tourist whose curiosity is by nature flitting and ephemeral, and who "merely regards the world as an object of the outer senses."[59] Such a tourist may well be content just to enjoy new sensations and exotic perceptions, and to collect a variety of souvenirs and anecdotes. The rationally curious traveler by contrast desires above all to see the world for the sake of "knowledge." Thus, whereas the tourist may wander at whim, the travelers who are rationally curious "want to profit from their journey." This means that they "must have a plan beforehand." In contrast to the physicist and the natural historian, both the tourist and the traveler do have this in common: for them the world is not just an object of knowledge but the stage upon which they move, the place where they live and journey. But whereas the prudent tourist makes up an itinerary for the sake of taking it all in, for the sake of knowledge the rationally curious traveler requires a "preconception" (*Vorbegriff*) of where he or she is going and the things he or she is setting out to see. It is in terms of this preconception that (beyond what direct or indirect enjoyment it may provide) the manifold of one's passing perceptions can be transformed into genuine experience and understanding. Analogously, then, scientific "travelers" seeking geographic knowledge of the world must also have a plan beforehand. They require a complete conception (*Inbegriff*) of geography "as a form of knowledge," in virtue of which then they can "anticipate future experience," their observations can be systematically ordered into a body of knowledge defined according to principles, and their knowledge can be properly "completed and corrected by experience." Such a conception is not itself an experience or an empirical datum, but an a priori condition of empirical *knowledge* in the area. It belongs then not to the body of empirical geographic knowledge but "to the *propædeutic* to our knowledge of the world."[60] Thus it does not in itself teach us how the world as a matter of fact is, but "how to know" the world as the object of outer sense, so that "within this complete conception we are in a position to show for every experience we have had, its class and place."[61]

Yet, despite its zetetic appeal, this metaphor of the traveler with his or her "plan" does not resolve or override other hesitations about the place of geography. Regarded as only a part of the doctrine of nature, the study of geography would seem to exemplify the sort of fragmentation of reason that works against rather than for the pursuit of worldly wisdom. Moreover, although it is true that all our actions take place in a spatial context, as we have seen it is not clear how for a strict deontologist like Kant geographic knowledge would be relevant to the question of moral worth and to that in which everyone "necessarily has an interest." But if geography is not relevant in this way, then it would be consigned to marginality. For as we have seen, in theoretical terms, it lacks centrality on formal foundational grounds. And although geographical knowledge has its instrumental value for builders, farmers, soldiers, and for everyone who needs to know the lay of the land in order to act prudently, it does not *directly* serve instrumental power in the way that other proper and improper natural sciences do. What is required then is a fuller account of the *value* of geography as knowledge. This issue might best be broached more directly not in terms of Kant's doctrine in the lectures but their pedagogical intent.

V

It is Kant's belief that for young children "the first lessons in science are most advantageously directed to geography," in particular to map work, since such study will discipline the child's "uncommonly strong imagination," and will help to develop the crucial power of memory.[62] In this form, the study of geography serves as a "good amusement" that nonetheless helps to develop habits needed for science and to broaden worldly horizons. At the university level, geographic study will help to remedy a different shortcoming. At this stage, Kant warns, the problem is "that youth learn *rational subtlety* early on, without possessing sufficient historical knowledge to be able to attend to the place of experience."[63] By the "place" of experience, Kant has in mind two things: the importance of experience to us as "citizens of the world" not as mere "artificers of reason," as well then as the place of experience as the actual stage upon which our worldly lives are played out. (Confronted with the contemporary phenomenon of philosophers who, much versed in the abstract subtleties of ethical metatheory and moral calculation, transmogrify themselves into "professional ethicists," yet without much appreciation for the actual experiences of those whom they would counsel, one might think that there is more than a youthful shortcoming to be remedied.)

As Kant presents it, the study of geography is also the means by which we first come scientifically to know (*machen bekannt*) nature *as a whole*, the epistemic goal being a "knowledge that is useful in all possible circumstances in life."[64] Understood in this way, geography comes before any more exact, proper, or improper natural science. It does so in the obvious way that geography treats of the same subject matter (viz., nature) as do such sciences, being scientific because it is systematic, and yet unlike the other natural sciences being readily accessible because it is popular. But in Kant's terms what is "popular" is not for that reason merely superficial, and hence something inevitably to be superseded. "For some time," Kant writes to a colleague, "I have been thinking . . . about the principles of popularity in the sciences . . . and I think that I have been able to define from this aspect not just another alternative [*Auswahl*] but an entirely different arrangement [*Ordnung*] from that demanded by the scholastic method."[65] Although offered as a general claim, this suggests that in the case of physical geography there is an appropriate "popular arrangement" in terms of which, according to its subjective origins at least, it can provide both the "compendium of nature" and a "purposeful arrangement of knowledge."

But if this popular arrangement is not just that of exact knowledge indiscriminately presented, then it must accord as well with the objective origins of geographic knowledge. In other words, for this "popular" arrangement to be methodologically legitimate, nature itself as the source and subject of such knowledge must be more than a mere aggregate of things under universal law. It must instead be a *world* of nature, to which then the compendium and the purposeful arrangement of knowledge properly conform. In this regard, Kant writes, "The word 'nature' assumes yet another meaning which defines the object. Whereas in the former sense it only denotes the *conformity to law* of the determinations of the existence of things in general, if we consider it *materialiter*, 'nature is the conceptual-complex [*Inbegriff*] of all the objects of experience.'"[66] In this latter characterization of nature, there is a direct link to nature as the object of geographic description. Kant writes, "We can view the world as the conceptual-complex [*Inbegriff*] of all knowledge of experience. The world as the object of outer sense is nature . . . [and] for knowledge of nature we are indebted to physical geography."[67]

Now, in strictly theoretical terms, knowledge of experience is for Kant simply knowledge "we obtain through the senses," the outer senses yielding knowledge of the world. The term "world" in this case coincides directly with the "substantive (*materialiter*)" definition of nature: it refers to "the conceptual-complex of all appearances."[68] Moreover, the sort of knowledge sought here is a knowledge of appearances as a conceptual-complex, that is, "of appearances insofar as, in virtue of an inner principle of causality, they are able to be thoroughly interconnected." It is in this sense then that geography is a doctrine of nature concerned with "objects of experience," and hence with finding facts and with determining concretely the "causes" of terrestrial phenomena.

In the preceding account, however, the central thread of Kant's argument is missing. To draw it out, we need to see how it is woven into the warp formed by two interrelated Kantian doctrines: First, Kant limits the range of genuine human knowledge to the range of possible experience. Within this range, we do have *genuine* a priori knowledge of nature, but only in the form of a rational knowledge of the a priori conditions of the possibility of experience. We cannot have any genuine "transcendent" knowledge of nature in itself, nor even legitimately seek to have such knowledge. For "finite rational being has nothing outside experience."[69] Hence if nature truly is a system in the (minimal) sense of a "conceptual-complex of objects of experience," this cannot be known of nature as a thing-in-itself. But it can be proved true of nature as "appearance," by being shown to be an a priori condition of the possibility of a pure science of nature, which means by being shown to be an a priori condition of the possibility of experience. Moreover, although this demonstration has to do with nature only as appearance, it establishes an *objective truth* about nature. For "in the world of sense, however deeply we inquire into objects, we have to do with nothing but appearances," whereas the thing-in-itself, albeit necessarily *thinkable* as the limit and ground (but not the cause) of our experience, is utterly, absolutely inaccessible to our *knowledge*, and so does not impugn that knowledge.[70]

Second, Kant argues for what in essence is a dialectical interrelation between the subject and the object of experience. In other words, he argues on the one hand that our existence as the conscious subjects of experience is not given, pace Descartes, in an act of sheer self-reflection, a pure *cogito ergo sum*, but that we first come to exist as conscious subjects only in relation to, and over against, an actually existent empirical reality (the refutation of idealism). On the other hand, he argues that this empirical reality is existent (i.e., "it is posited"[71]) only as it is constituted as such by and for a subject in the form of *objects* of experience, part of a *system* of objects. Thus, "the conditions of the possibility of experience in general are likewise conditions of the possibility of the objects of experience," and the essential possibility of objective judgments of experience in general comes before that of our merely subjective judgments of perception.[72]

These two doctrines provide the context in terms of which Kant proceeds from the abstract, formal definition of nature as the existence of things in general under universal law to the definition of nature *materialiter* "as the conceptual-complex of appearances." He makes this move by demonstrating that the latter conception of nature is a necessary condition for the very possibility of experience in general, and hence that, along with the conformity of things to law, this conception is required too by natural science proper. In strictly theoretical terms, then, the science of geography as an empirical doctrine of nature, and Newtonian physics, as Kant's paradigmatic natural science, *both* treat of nature as "objects of experience" understood as a "conceptual-complex of appearances." Now, it would be a task for scholars of Kant's philosophy of science to determine precisely how in significant respects geography and empirical physics would then differ as research approaches. Suffice it here to point out the following themes more or less implicit in Kant's account.

First, although physics and geography both treat of the same object (i.e., nature as a "conceptual complex of appearances"), they do so at different removes from the knowledge of that object as a knowledge of experience *in life*. Geography, as we have said, is the means by which we first come *scientifically* to know nature *as a whole*, providing the "idea" in terms of which our everyday experience of objects in and as a world can first be properly ordered and developed into a rigorous system of knowledge, that is, knowledge ordered explicitly by principles and in principle objectively verifiable. The research of empirical physics is yet at a further remove from that of geography, and thus twice removed from the knowledge of experience in life. The "conceptual-complex of appearances" that is the object of physics is essentially an ideal construct. In other words, it is not an arrangement that physics simply takes over from everyday experience and refines according to a rational idea or principle (i.e., what appropriately geography in its "popularity" does). Rather, as Heidegger points out, it is a domain of objects that methodologically physics first *posits*, creating thereby a fixed ground-plan (*Grundriß*) of natural events in terms of which physics-research proceeds.[73] In this respect, then, physics is more "exact" than geography precisely because it is more properly a "positive/positing" science. In other words, it "is not exact because it calculates with precision; rather it must calculate in this way because its adherence to its object-sphere has the character of exactitude. The humanistic sciences, in contrast, indeed all the sciences concerned with life, must necessarily be inexact in order to remain rigorous."[74]

Insofar as geography is more immediately related to knowledge of the world in life, it provides the general framework within which *all* our knowledge of nature is to be ordered. It does so in the sense that it serves as the middle-term between the common knowledge of the world we have in life, whose guiding principles are not critically demonstrated, and the methodical positing and precise mathematical calculation of the world of nature that is the work of physics. Kant does maintain on the one hand that "in any particular doctrine of nature only so much genuine science can be found as there is mathematics to be found in it."[75] Newtonian physics thus remains the paradigmatic natural science, in relation to which geography as a doctrine of nature falls short. But on the other hand, this does not diminish essentially the role of geography. For in its essence, geography is not simply nor ultimately a doctrine of nature, but rather a "world-knowledge." In this characterization, moreover, "world" does not mean nature *materialiter* as an objective complex of appearances, but the world as a *pragmatic* context of meaning. In this sense, the world is not principally the object of theoretical and scientific calculation and dominion, but the place of our pragmatic domicile. Likewise, the things of the world are not objects first posited by scientific research, but those things with which first and foremost we have to deal and to which we relate purposively (i.e., the *pragmata* of our everyday *praxis*[76]), the things then in terms of which we situate ourselves. In its essence, as one commentator observes, "geography is a science of the earth as the location [*Standort*] of humanity: It reveals the world not only as 'objective', but also essentially as the *meaningful* world for humanity, which as such 'belongs to us' so that in it we have 'our place'."[77]

This essential characterization of geography points to the philosophical value of geographic study. To see this, however, we need to recall in Kantian terms the value of philosophy itself. Like geography, philosophy too has not only a scholastic, theoretical meaning, but also a "cosmological" practical one. "According to its world-concept," Kant writes, "philosophy is the science of the ultimate ends of human reason," and therefore has to do with the whole vocation (*Bestimmung*) of humanity, with how we are to fulfill our proper place in the scheme of things and understand what we must do in order to be truly human.[78] Moreover, not only does "this high concept give to philosophy its *dignity*, that is, an absolute worth," but philosophy conceived in this way also in the end "gives value to all other cognitions." It is from this perspective that geography, as a "propædeutic to our knowledge of the world," receives its true value.

VI

Yet, rather than exhibiting a coherent arrangement of the parts of the Kantian system, the preceding account of the place of geography points to a fundamental tension in Kant's thought. On the one hand, insofar as Kant's turn to geography subtends the perspective of the positive sciences, in which the world is simply the object of knowledge, with the more original pragmatic account of the world as the context of our life, "it discovers in principle an entirely new foundation [*Grundstellung*] of science, just as it discovers a new horizon of philosophy."[79] On the other hand, the requirements of the critical system preclude Kant himself from realizing the full implications of this discovery. This is quite obvious in the theoretical domain. Despite the importance accorded to geography with respect to the foundations of science, it does not impugn the paradigmatic importance of the exact sciences. This points to a fundamental problem. For it is not clear how geography can both truly provide the general framework for our knowledge of nature, and thus serve as a propædeutic to physics, yet in its own scientific procedures fall short of the paradigm. Similarly, between the scientific status of geography as an empirical doctrine of nature, and geography understood essentially from a pragmatic viewpoint, there is no obvious, easy fit. In this respect, the "idea" of geography would seem to be two, rather than one.

In the practical domain, there are similar problems. Geography is essentially "world-knowledge" understood pragmatically, rather than theoretically. "Pragmatic knowledge investigates what *humanity* as a free agent makes, can make or should make of itself," which is different from the sort of "physiological" knowledge that concerns what "nature makes of humanity."[80] But it is also different from practical, moral knowledge, and from what humanity does *essentially* to realize its nature. To be sure, the pragmatic perspective concerns what humanity "should" make of itself, but this "should" is not strictly a moral one. The moral "should" is categorical, whereas the pragmatic is hypothetical. The latter has to do with what in general humanity "as an animal endowed with the capacity of reason (*animal rationabilis*) can make of itself as a rational animal (*animal rationale*).''[81] Accordingly, the pragmatic "should" says: "If, in this or that specific circumstance, one wants to realize one's rational capacity, then one *should* act in certain ways." In contrast, the moral "should" says: "Regardless of circumstances, one should will the moral law." Now in two respects this doctrine does grant considerable importance to the pragmatic. First, since the pragmatic has to do with how we, as beings who by nature are *animal rationabilis*, make ourselves rational by realizing the rational potential grounded in our nature, the force of its "should" is not purely hypothetical. For in general our rational self-determination is the source of our human dignity, and hence the command to act "rationally" in life is more than a matter of sheer prudence and skill. Second, precisely because qua human we are finite and conditioned, whereas qua rational our duties are absolute, the most perplexing dilemmas for the morally worthy are not the choices between duty and heteronomous ends, but the fateful choice between two equally binding duties. In such circumstances, then, considerations of the *bonum pragmaticum*, although not determining moral worth essentially, might well be relevant to the actual choice made. On the other hand, this doctrine effectively takes away any *ultimate* significance for the pragmatic. Although the pragmatic perspective speaks of humanity "making" itself, this is not an ontologically decisive self-making. Indeed, for Kant (at least in the published texts), the claim is that "human being does not . . . make itself."[82] Our dignity lies in our rationality; but our distinctive rationality, that reason in and through which we truly fulfill our being, doing what we must do in order to be truly human, is moral reason. Hence it is not that reason which, skilled in the ways of the world, confirms itself by its worldly results, but that reason which, in recognizing and acting upon an unconditioned moral law, realizes and confirms itself purely in terms of itself, and is "awed" by the prospect.

VII

At this point, I cannot presume to resolve the tensions I have noted in Kant's doctrine, nor then go on to develop and defend fully the alternative thesis implied by his "discovery." Suffice it instead merely to indicate the sorts of assumptions that would be required to resolve the tensions I have noted keeping to the trajectory that Kant himself sets, and to state what broad categories would be required in order to make intelligible the thesis of the "ethical determination of geography" that emerges from these assumptions. With respect to the first task, the work is made easier by the fact that Kant's immediate successors, in particular Fichte, already make explicit the assumptions I have in mind. The relevant doctrines may be stated as follows:

1. The only intelligible and defensible thesis regarding human selfhood is that the self is essentially a self-constituting process, that "a human being is nothing but the series of its acts."[83] In one form or another, this thesis has, until recently, informed the entire mainstream of post-Kantian continental thought.[84]

2. In and through this self-constituting process, human beings at the same time make actual in the world the transcending intelligibility, rationality, or meaning that accounts for the existence of that world as such. Put in negative terms, this doctrine speaks against the notion of an abstract intelligibility or timeless reason, fully actual in itself, which human beings then simply recognize and act upon in particular cases, thereby realizing a potential grounded in a permanent human nature. Put in positive terms, it means that the transcending intelligibility,

meaning, or reason is only truly actual insofar as it is realized (in both senses of the English term, i.e., recognized and made real) in and through our self-constitution. To illustrate: For Kant, when one acts in accord with the moral law, one recognizes and gives this law to oneself, the law being valid unconditionally whether I acknowledge it or not. By contrast, Fichte argues that in moral action the self-legislating self *produces* the law in virtue of realizing it, and so has a share in the creation of the intelligible world. The moral law is still rational and universal; but the actuality of the rational depends upon our making it actual in and through our self-making in accord with it.

3. Insofar as the self is self-constituting essentially, then the division between our physiological nature, or what nature makes of us, and what we make of ourselves (understood here not just pragmatically but existential/ontologically) is subject to essential mediation. Thus, for example, Kant insists on a sharp untranscendable division in human being itself between our being as morally free and thus unqualifiedly responsible, and our being as subject to unalterably given natural inclinations and hence as innocent. By contrast, Fichte regards the inclinations, not as natural givens, but as being actual as such only in relation to our moral self-legislation, in and through which, moreover, they are subject to definitive transformation and gradual elimination (*Aufhebung*) as factors capable of giving absolute resistance to the moral will.

4. Insofar as human selfhood is not divine, and hence does not, as it were, produce it itself *ex nihilo*, its self-making takes place within essential, situating conditions (i.e., within the limits of nature, history, and the human condition as a totality). The relation between the self and its situating limits is dialectical and ontologically decisive. According to Fichte, for example, our self-making lies in our moral self-determination, in relation to which the world of nature is "the material of my duty rendered sensuous."[85] In this view, we *are* our moral self-legislation, and the world *is* the material of our duty; the dialectic encompasses the very *being* of both terms as inextricably related.

Of course, much more would need to be said to make these doctrines fully intelligible and defensible. Still, enough has been sketched to enable us to identify in terms of them the root of the problem in Kant's project, and to see in principle a remedy. However much Kant valorizes geographic knowledge philosophically, he cannot in the end overcome the freedom/nature opposition that would be required in order to legitimate that valorization. He is left then on the one hand with a subject defined essentially in terms of unconditioned moral freedom, and on the other with a nature (including our empirical nature) defined essentially as a complex of objective appearances determined according to an inner causal principle. The "cosmological" understanding that might seem to mediate this opposition fails, precisely because it does not truly encompass the *essence* of either realm. With respect to geography, Kant recognizes two perspectives—the world as an objective natural environment for geography taken as a doctrine of nature, and the world as the pragmatic stage for geography taken as world-knowledge—and he clearly ranks the latter above the former in principle. But his emphasis on the pragmatic and on the worldly in this latter sense does not serve to join together what critical reason has already put asunder. For the pragmatic and worldly are not essential *theoretically* to how transcendentally we constitute experience, nor are they *essential* practically to the person that, according to moral reason, we determine ourselves to be. With respect then to Kant's conception of geography, the particular problem is this: To the extent that the worldly and pragmatic perspective that defines geography essentially is analogous to the nature/freedom division, it does not constitute a distinctive third term that could mediate that division; and insofar as it does constitute a distinctive perspective different from the theoretical and the practical viewpoints, it merely stands juxtaposed to these two domains, being not their synthesis, but simply *another* perspective.

This points to the root of the problem. Kant does affirm that *reason* is essentially a *self*-activity. In other words, "it *is* reason only when it is *self*-determined," recognizing in nature on the one hand only what reason has "produced after a plan of its own," and recognizing in the moral will on the other only what autonomously it has given to itself.[86] Yet, for Kant, these two realms of reason's self-activity remain abstractly related. To be sure, Kant does recognize that in our human being the self-activity of reason is not the activity of two natures *realiter*, but rather represents the activity of the one human nature viewed from two perspectives. But he cannot show how these two perspectives are truly united *in knowledge*, nor then how in the world of experience these two realms of reason's self-activity are truly one. He cannot because he does not grant that our human being is a self-constituting process. If he did, then he would be led to conclude (as were his idealist successors) that no joining together of the two dimensions of reason's self-activity is required, since these dimensions would always already be joined in the self-constituting process that *sui generis* establishes one's self-identity in the first place.

Thus, likewise, although Kant gestures toward the importance of the pragmatic and the worldly with respect to geography, ultimately in this gesture the two dimensions (i.e., the pragmatic activity of the subject and the world as the place of that activity) remain abstractly juxtaposed. They are two dimensions that are only *viewed as one* through a quasi-existential lens; yet in effect they remain two. Kant cannot show that they have an intrinsic relation, because he does not understand their relation existentially. In other words, he does not recognize on the one hand that our pragmatic activity consists not just in how we happen to deal with things, but that it *is* the ontological process of our self-constitution. On the other hand, he does not recognize that the world is more than an objective natural environment wherein we happen to live and deal with things, but is the essential situating context of our self-constitution. Viewed from this angle, the root of the problem is that Kant does not fully appreciate the dialectic that governs the relation of human being and natural world.

To appreciate this dialectic fully is to see that there is not first a given self who, as an agent, then comes to the world with particular plans, projects, and strategies; nor that there is first a given world that is determined in itself as nature and laid out before us as an objective fact, and then as such serves as the "stage and substrate" upon which the agent acts. In any dialectical relation, the terms of the relation are equiprimordial, differentiated yet mutually dependent. In this particular case, the natural world, which situates the self-constituting self, must be other than self—otherwise it would not serve to situate the self. Yet, precisely as situating the self, the natural world must enter into the self's self-constitution essentially, and thus not be *wholly other*.[87] Otherwise, in this latter respect, the self would either be a mere product of the natural world (as materialism in all its guises claims), and hence not a thing essentially different from other worldly things. The question of the relation of human being and natural world would then only be a matter of the empirical calculation among matters of fact. Or the self would be given as such entirely independent of the world (as rationalism in all its guises claims), and hence in its being would be related to world, if at all, only externally and accidentally. The question of the relation of human being and the natural world would thus be, philosophically, a matter of indifference. But in the dialectic of human being and natural world, the finite human self is what and how it is only as it is self-constituting in relation to the natural world not as a sheer objective fact but as a situating context that limits my selfhood. In this relation, moreover, the notion of "limit" is itself dialectical: It not only denotes an otherness that restricts my selfhood qua finite, but also is a positive boundary within which the self comes to be a self in the first place. Moreover, to speak of this dialectic as a relation of human being and nature is already a falsifying abstraction. At one and the same time, the meaning and reality of nature itself is a function of our self-constituting acts, as the meaning and reality of our human being is a function of the natural situation. In this dialectic, then, both the external relation of history and geography, and the Kantian priority accorded to geography over history are always already overcome.

VIII

It remains to state—and at this point, no more than to state—how such changes rung on the Kantian perspective yield in principle an ethical determination of geography. It is important first to remind ourselves of two Kantian insights, both skeptical, which inform the perspective being sketched. First, Kant wisely rejects any notion of establishing our relation to nature in terms of some presumed "hyperphysical" insight into nature as such, from which we then adduce ethical and pragmatic precepts. He would rightly regard such speculations as a mystification. Hence he would likely reject contemporary strategies that attempt to derive an environmental ethics from some presumed insight into the metaphysical unity of nature, supposed then to yield normative precepts about how one ought to experience and interact with nature. Second, even if Kant were to accept the four doctrines I have sketched, he would still likely reject the sort of ethical idealism (identified with Fichte) that was first adduced from them. Kant does concede that it is an intrinsic tendency of our reason to regard the natural world as a system of purposes arranged with humanity as a final purpose. But he is rightly sarcastic about the sorts of conclusions that one might be tempted to draw, if one were to take this tendency to correspond to objective truth.[88] Thus, Kant would object to Fichte's attempt to see the world as a limit upon the self only ever as the "self's *self*-limitation," and hence to consider the natural world as "incapable of giving absolute resistance to the moral will."[89] A successful idealism would have to demonstrate "the ideality of the finite" (from Hegel) and hence show through all experience that the self, although dialectically related to nature, is yet capable of transcending natural limitations. Fichte's ethical idealism simply denies *ex hypothesi* that nature is ever genuinely other-than-self.

The notion of an ethical determination of geography properly makes sense, only if we understand it in an existential, ontological way. To put the matter abstractly: (1) It is intelligible and defensible only if the "geographic" refers, not to a "description of the earth" where "earth" is taken to mean the natural environment given in everyday experience and with which we must deal, but to a subset of the situating natural conditions of our being-in-the-world. Such conditions are both natural and conditions only in relation to our self-constitution. Natural conditions are those situating capacities and structures that do not originate directly from human artifice, the geographic being the subset of those capacities and structures not belonging to the particular body that one's self, as a center of initiative, has at its disposal. (2) "Determinations" refers here not to objective causal factors in the natural environment that have an objective effect on human life. If we look at matters in this way, "we find ourselves examining the relation between object and object," and not between self and situation.[90] Instead, "determination" is to be understood in terms of our essential self-constitution, determinations being situating factors of the self that are determinative "only as they belong to the structure of human existence," that is to its self-constitution. (3) In this usage, the ethical does not refer to rules, principles, or precepts that a human agent is to follow in dealing with the world. Rather "in keeping with the basic meaning of the word '*ethos*', we should now say that 'ethics' ponders the abode of humanity."[91] To be sure, ethics in this sense offers nothing like a traditional moral philosophy, be it based on *ideas*, natures, an old or a new categorical imperative. Nor then does it issue in specific injunctions concerning how we ought to treat the environment. The appeal to the ethical here is an appeal to explore the essential-space (*Wesensraum*) of our being-in-the-world, as the true "substratum and stage" for all of our particular cognitions, actions, and skills, and the true "ground" from which our theoretical and practical reasoning "arises and is applied." The exploration of this essential-space allows us to recognize and appreciate where concretely we truly stand in the world, as well as to help to recognize and appreciate the full dimensions of all that to which we are indebted for our being. Such thinking does not issue in knowledge. But, as Kant himself recognized, "our pressing need . . . is more than the mere quest and desire for knowledge."[92] Our reason seeks ultimate integration, the sense of things as such and as a whole. The thinking that seeks to discover that sense is an original ethics.

Notes

1. Clarence J. Glacken, *Traces on the Rhodian Shore* (Berkeley and Los Angeles: University of California Press, 1967), xiii.
2. David N. Livingstone, *The Geographical Tradition: Episodes in the History of a Contested Enterprise* (Oxford: Blackwell, 1992), 28.
3. Jean-François Lyotard, *La condition postmoderne* (Paris: Éditions de minuit, 1979), 7.
4. Martin Heidegger, *Kant und das Problem der Metaphysik,* 4 Aufl. (Frankfurt: Klostermann, 1973), 198.
5. Preston E. James, *Encyclopeadia Britannica,* 1968 ed., s.v. "Geography"; George Tatham, "Geography in the Nineteenth Century," in *Geography in the Twentieth Century,* ed. Griffith Taylor (London: Methuen, 1957), 38; and Jan O. M. Broek, *Geography: Its Scope and Spirit* (Columbus: Charles E. Merrill, 1965), 14. (All quoted in J. A. May, *Kant's Concept of Geography* [Toronto: University of Toronto Press, 1970], 8–9.)
6. See, e.g., Alfred Hettner, *Die Geographie: ihre Geschichte, ihr Wesen und ihre Methoden* (Breslau: Ferdinand Hirt, 1927), 447.
7. Paul Gedan, "Immanuel Kants Physische Geographie," in *Kants Gesammelte Schriften* (Berlin: Prussian Academy, 1902-38), 9: 509-10. In that second semester, Kant also adds a course on the foundations of general natural science (see Ernst Cassirer, *Kant's Life and Thought,* trans. James Haden [New Haven: Yale University Press, 1981], 41).
8. May, *Kant's Concept,* 4. May also writes, "At Göttingen in 1754-55, Büsching had introduced courses on the globe and the political geography of Europe; and in the following year, J. M. Franz taught a course on the geography of North America."
9. An example of a philosopher who regarded geographic knowledge as part of general enlightenment and worldliness is Locke, who recommends geography as an essential part of a "gentleman's" self-improvement, a knowledge of geography being "absolutely necessary . . . to the reading of history," and required in order to "well understand a Gazette" ("Some Thoughts concerning Reading and Study for a Gentleman," *The Works of John Locke* [London: Thomas Tegg, 1823], 3: 297–98). This is a point that Kant reiterates ("Physische Geographie," *Kants Gesammelte Schriften,* 9: 163).

In regard to others who had included geography in a revised science, the notable examples are more numerous. In the *Parasceve,* for instance, Francis Bacon not only mentions here and there what we would nowadays regard as geographic topics, but also in the appended "catalogue of particular histories," he specifically lists *Historia Geographia Naturalis,* being the "natural history

... of mountains, valleys, woodlands, plains, sands, marshes, lakes, rivers, torrents, springs, and every variety of their course, and the like." He thereby assigns to geography a definite place within the *Novis orbis scientiarum*, though precisely as a *natural* history that "pass[es] over nations, provinces, cities, and such like matters pertaining to civil life" (*The Works of Francis Bacon*, ed. James Spedding, Robert Leslie Ellis, and Douglas Denon Heath [Boston: Houghton Mifflin, 1857], 2: 42–69, esp. 62–63). Likewise, in setting forth "the several subjects of knowledge," Hobbes includes geography among the sciences that make up "natural philosophy," naming the subject, together with astromony, as the species of cosmography, that is, the science that calculates "consequences from the motions and quantity of the greater parts of the world, as the earth and the stars." Now, it is Hobbes's view that knowledge as such divides into two broad classes—"history" as "knowledge of facts," being either "natural" or "civil history," and "science/philosophy" as "knowledge of consequences," being either "politiques and civil philosophy" or "natural philosophy." Although he assigns geography a definite place in this last division, whether it also has a place among civil and/or natural history, Hobbes does not say (*Leviathan*, ed. C. B. MacPherson [Harmondsworth: Penguin, 1968], 147–49). By contrast, in Vico's *Scienza nuova* it is precisely the "civil" origin of geography that is brought to the fore. This is understandable, since Vico's avowed concern is not "the study of the world of nature, which, since God made it, He alone knows," but the "study of the world of nations, or civil world, which, since men had made it, men could come to know" (*The New Science of Giambattista Vico*, trans. T. G. Bergin and Max Fisch [Ithaca: Cornell University Press, 1984], 96 [§331]). From this perspective, Vico claims for his new science "a discovery . . . which is just the opposite of that of Bacon" (Bergin and Fisch, xli). For whereas Bacon "considers how the sciences as they now stand may be carried on toward perfection, this work of ours discovers the ancient world of the sciences, how rough they had to be at birth, and how gradually refined, until they have reached the form in which we have received them." It is in the context of recalling the poetic beginnings of the arts and sciences that Vico considers how the Greeks had "created" geographic learning, and "how by geography [they] . . . had described the whole world within their own Greece" (Bergin and Fisch, 112 [§367], and 285–96 [§741–78]). Less eccentric in this regard, D'Alembert claims that, along with chronology, geography is an "offshoot and support" of "history"—the latter being the science that "locates peoples in time"; the former, the science of "place" that "distrubutes [peoples] over the globe" (*Preliminary Discourse to the Encyclopædia of Diderot*, trans. Richard Schwab and Walter Rex [New York: Bobbs-Merrill, 1963], 144–45 and 153–54). Thus, unlike Bacon and (perhaps) Hobbes, D'Alembert thinks that geography "draws considerable help . . . from historical facts," and distinguishes the subject from "geology" as that branch of physics that inquires into the causes of physical terrestrial phenomena. Nonetheless, he still defines geography nonhumanistically as a "mathematical description of the earth," and insists on its peculiar fact-finding character.

For examples of those who regarded geography as important in a general way, Jean Bodin claims that the natural inclinations of peoples are conditioned by their physical environment, making it incumbent on a wise legislator to understand such relations sufficiently well in order then to "know when and how to overcome, and when and how to humor, these inclinations" (*Six Books of the Commonwealth,* trans. and abridged by M. J. Tooley [Oxford: Blackwell's, 1955], 157). In a similar vein, Montesquieu devotes five books of *L'esprit de lois* to discussion of the relation of laws to climate and soil. An additional, though less striking example, is found in Rousseau—the philosopher whom Kant celebrates as having "first discovered . . . humanity's deeply hidden nature" ("Bemerkungen zu den Beobactungen," *Kants Gesammelte Schriften,* 20: 58). In asking "whether there can be some legitimate and sure rule of adminsitration in the civil order," Rousseau reasons that, "just as nature has set limits to the status of the well-formed man . . . so too with regard to the best constituton of the state, there are limits to the size it can have," and goes on to claim that for the best constitution the optimum relation of territory to population must take account of the "idiosyncrasies of a place" (e.g., that mountain women "are more fertile than those on the plains" [*Du contrat social* (Paris: Flammarion, 1992), 27, 71 and 74]). Among Kant's contemporaries are the examples of Winckelman, who affirms an intrinsic connection between climate, health, and the ideal of beauty (*A History of Art among the Greeks,* trans. G. Henry Lodge [London: John Chapman, 1850], 5), and of Herder, who attempts more ambitiously to work out the rudiments of a theory of an organic relation between humanity and the physical environment (see, e.g., F. M. Barnard, *Herder's Social and Political Thought* [Oxford: Clarendon Press, 1965], 121–22).

10. "Physische Geographie," 156.

11. May, *Kant's Concept,* 4.

12. . Kant, "Entwurf u. Ankündigung eines Collegii der physischen Geographie," *Kants Gesammelte Schriften,* 2: 25 and 35.

13. See Friedrich Paulson, *Immanuel Kant: His Life and Doctrine,* trans. J. E. Creighton and Albert Lefevre (New York: Ungar, 1963), 60.

14. He had, for example, to sell off his personal library over the years just to make ends meet (Drescher, hrsg., *Wer was Kant?,* 137). See also, Briefwechsel, 10, 19 (To Lindner, 28/October/1759).

15. See "Anthropologie in pragmatischer Hinsicht," 7: 122, for popular literature; see, also, Cassirer, *Kant's Life and Thought,* 40. It is a well-known piece of Kantian lore that throughout his life he read works of geography as his chief means of mental relaxation (see, Willibalde Klinke, *Kant for Everyman,* trans. M. Bullock [London: Routledge and Kegan Paul, 1952], 22). It is also evident that he was sensitive to geographic concerns, for example: that he extolled the virtues of his native Königsberg for "broadening one's

knowledge of humanity and the world . . . without traveling" ("Anthropologie in pragmatischer Hinsicht," 120–21); and that this attachment, together with a "temperament that cannot resolve to make changes that to others seem only trifling," led him in later years to decline lucrative offers from other universities (see, e.g., Briefwechsel, 10, 83 [To Suckow, 15/12/1769]). On this latter score, it might be worth comparing Kant's response to such offers with Heidegger's pretentious, *volkstümlich* response ("Schöperische Landschaft: Warum bleiben wir in der Provinz?" *Denkerfahrungen* [Frankfurt: Klostermann, 1983], 9–13). For "entertainment and conversation," see "Physische Geographie," 165. Kant's "new direction" in philosophy is discussed in "Prolegomena zu einer jeden künftigen Metaphysik," *Kants Gesammelte Schriften*, 4: 260.

16. "Logik," *Kants Gesammelte Schriften*, 9: 23.
17. "Logik," 9: 22; Cf. Kant, *Vorlesungen über die Metaphysik.* Hrsg. K. H. Schmidt (Roteswein: J.H. Pflugbeil, 1924), 1.
18. "Geography is a science which teacheth the description of the whole earth" (N. Carpenter, *Geographie delineated forth in two Bookes* [Oxford: Oxford University Press, 1695], 2); see also "Physische Geographie," 157.
19. "Physische Geographie," 160.
20. Aristotle, *Metaphysics,* 1026b. On common knowledge and science see "Logik," *Kants Gesammelte Schriften* 9: 72.
21. Kant, *The Critique of Pure Reason,* trans. N. K. Smith (New York: St. Martin's, 1929), 653 (B 860).
22. "Physische Geographie," 158.
23. "Physische Geographie," 164.
24. "Physische Geographie," 158. Cf. also "Physische Geographie," 156–57 and "Anthropologie in pragmatischer Hinsicht," *Kant's Gesammelte Schriften,* 7: 119–20.
25. "Physische Geographie," 160 and 162 (my emphasis).
26. "Die metaphysische Anfansgründe der Naturwissenschaft," *Kant's Gesammelte Schriften,* 4: 468.
27. "Physische Geographie," 161.
28. "Physische Geographie," 157.
29. "Physische Geographie," 162.
30. "Physische Geographie," 163.
31. "Die metaphysische Anfansgründe der Naturwissenschaft," 469.
32. "Die metaphysische Anfansgründe der Naturwissenschaft," 467n.
33. "Die metaphysische Anfansgründe der Naturwissenschaft," 468.
34. "Physische Geographie," 198–221 passim, 239–40, 260–63.
35. "Prolegomena zu einer jeden künftigen Metaphysik," 294.
36. See, e.g., *The Critique of Pure Reason,* 162 (B 147).
37. Kant, *Critique of Judgment,* trans. W.S. Pluhar (Indianapolis: Hackett, 1987), 23.
38. *Critique of Judgment,* 302 (§79).

39. E. Husserl, *The Crisis of European Sciences and Transcendental Phenomenology*, trans. David Carr (Evanston: Northwestern University Press, 1970), 48–49.
40. *The Critique of Pure Reason*, 20 (B xiii–xiv).
41. "Physische Geographie," 165.
42. "Grundlegung zur Metaphysik der Sitten," *Kant's Gesammelte Schriften*, 4: 408 and 431.
43. "Grundlegung zur Metaphysik der Sitten," 431
44. "Grundlegung zur Metaphysik der Sitten," 428 (my emphasis).
45. "Freedom is certainly the *ratio essendi* of the moral law, but the moral law is the *ratio cognoscendi* of freedom" (*The Critique of Practical Reason*, trans. L. W. Beck [New York: Macmillan, 1993], 4).
46. "Grundlegung zur Metaphysik der Sitten," 411.
47. Hans Jonas, *The Imperative of Responsibility: In Search of an Ethics for the Age of Technology* (Chicago: University of Chicago Press, 1984), 12.
48. *The Critique of Pure Reason*, 29 (B xxx), my translation.
49. "Bemerkungen zu den Beobactungen," 44.
50. *The Critique of Pure Reason*, 4 (B ii).
51. See, e.g., Kant, "What Is Enlightenment?" in *On History*, ed. L.W. Beck (New York: Bobbs-Merrill, 1963), and *The Conflict of the Faculties*, trans. Mary J. Gregor (Lincoln: University of Nebraska Press, 1992), 13ff.
52. Kant, "Nachricht von der Einrichtung seiner Vorlesungen," *Kant's Gesammelte Schriften*, 2: 305.
53. "Nachricht von der Einrichtung seiner Vorlesungen," 310.
54. *The Critique of Pure Reason*, 657–58 (B 866–68); my emphasis.
55. Kant, "Über das mißlingen aller philosophischen Versuche in der Theodizee," *Kant's Gesammelte Schriften*, 8: 197.
56. "Anthropologie in pragmatischer Hinsicht," 199.
57. "Nachricht von der Einrichtung seiner Vorlesungen," 307.
58. Kant, "Entwurf eines Collegii der physischen Geographie," 3.
59. "Physische Geographie," 157.
60. "Physische Geographie," 157 (my emphasis).
61. "Physische Geographie," 158.
62. Kant, "Pädagogik," *Kant's Gesammelte Schriften*, 9: 474–76.
63. Kant, "Nachricht von der Einrichtung seiner Vorlesungen," 312.
64. "Nachricht von der Einrichtung seiner Vorlesungen," 157.
65. "Briefwechsel," 10: 247 (To Herz, January 1779).
66. "Prolegomena zu einer jeden künftigen Metaphysik," 295.
67. "Physische Geographie," 156–57.
68. *The Critique of Pure Reason*, 392–93 (B 446–47).
69. This is Fichte's formulation (*Werke*, Hrsg. I. H. Fichte [Berlin: De Gruyter, 1971], 1: 425).
70. *The Critique of Pure Reason*, 84 (A 45).
71. *The Critique of Pure Reason*, 504 (B 626).

72. *The Critique of Pure Reason,* 194 (B 197).
73. M. Heidegger, "Die Zeit des Weltbildes," *Holzwege* (Frankfurt: Klostermann, 1972), 71ff.
74. "Die Zeit des Weltbildes," 73.
75. "Die metaphysischen Anfangesgründe der Naturwissenschaft," 4: 470.
76. Cf. M. Heidegger, *Being and Time,* trans. J. Macquarrie and E. Robinson (New York: Harper & Row, 1962), 96–97 (§15).
77. G. Krüger, *Philosophie und Moral in der Kantischen Kritik* (Tübingen: J.C.B. Mohr, 1931), 41.
78. "Logik," *Kant's Gesammelte Schriften,* 9: 23.
79. Krüger, *Philosophie und Moral in der Kantischen Kritik,* 41.
80. "Anthropologie in pragmatischer Hinsicht," 119.
81. "Anthropologie in pragmatischer Hinsicht," 321.
82. In the *Opus Posthumum,* however, Kant does speak of the "self-determining subject," and of "making onself" (*Opus Posthumum,* trans. E. Föster and M. Rosen [Cambrige: Cambridge University Press, 1993], 254). See also "Grundlegung zur Metaphysik der Sitten," 451.
83. G. W. F. Hegel, *The Encyclopædia Logic,* trans. T. F. Geræts, W. A. Suchting and H. S. Harris (Indianapolis: Hackeet, 1991), 212 (§140).
84. Some salient examples being: "The ego posits itself" (Fichte); "I am because I am" (Schelling); "Human being is what it does" (Hegel); "Human being is [ist] what it eats [ißt]" (Feuerbach); "Human being is the product of its labor" (Marx); "I choose myself in my eternal validity" (Kierkegaard); "I am what I will" (Nietzsche); "Existence precedes essence" (Sartre); "The essence of Dasein lies in its existence" (Heidegger). For a clear and profound exploration of this thesis, see, E. L. Fackenheim, *Metaphysics and Historicity* (Milwaukee: Marquette University Press, 1961).
85. Fichte, *Werke,* 5: 185.
86. E. L. Fackenheim, *The Religious Dimension in Hegel's Thought* (Boston: Beacon Press, 1967), 225.
87. See, Fackenheim, *Metaphysics and Historicity,* 44ff.
88. Thus, for example, from the arrangement of geographic conditons that permit human beings to survive in colds lands, one might infer that nature itself serves the purpose of such survival. Yet to infer this, Kant says, is to miss the crucial issue, namely, that of identifying and remedying the social disunity that "could have dispersed people into such inhospitable regions" in the first place (*Critique of Judgment,* 247 [§63]).
89. Fackenheim, *Metaphysics and Historicity,* 45–46.
90. Watsuji Tetsuro, *Fudo: Wind und Erde,* Ubers, Dora Fischer-Barnicol u. Okochi Ryogi (Darmstadt: Wissenschaftsbuchgesellschaft, 1992), 1.
91. M. Heidegger, "Letter on Humanism," *Basic Writings,* trans. D. Farrell Krell (New York: Harper & Row, 1977), 234–35.
92. "Prolegomena zu einer jeden künftigen Metaphysik," 368.

Nature's Presence: Reflections on Healing and Domination

Eric Katz

I

The trees are like a forest. Although I can hear the sounds of traffic on Okopowa Street on the other side of the wall, inside the Jewish Cemetery of Warsaw all is quiet. There is a light rain and fog. In the grayness of the day, the mist and the shadows prevent my eyes from seeing deep into the cemetery. All I can see are the trees and the underbrush, lush and green, growing up and over the scattered and crooked grave stones. One main walkway and a few paths have been cleared, so that tourists can view several hundred of the tombstones. Another path leads to a clearing. It is a clearing of tombstones, not of trees, the mass grave of the Jews who died in the Warsaw Ghetto before the deportations to Treblinka began in July 1942. The mass grave appears as a meadow under a canopy of tree branches. The area is ringed by grave stones, but the center of the clearing is covered with grass. Dozens of memorial candles flicker, remaining lit despite the dampness and the light rain. The beauty of this mass grave surprises and shocks me. Here is the reification of irony. This cemetery, a monument to the destructive hatred of the Nazi Holocaust, is extraordinarily beautiful. Filled with a vibrant, unchecked growth of trees and other vegetation, the cemetery demonstrates the power of Nature to re-assert itself in the midst of human destruction and human evil.

The next day I travel to Lublin, near the Ukrainian border—a two-hour drive from Warsaw, through endless flat farmland where Polish farmers still use horses to plow the fields. It is harvest season, and the car slows occasionally to pass a truck filled with sugar beets. Our destination is Majdanek, the death camp lying three kilometers from the center of Lublin. Majdanek fills a treeless meadow stretching as far as the eye can see. Standing at the entrance gate one can see in the distance, a mile off, the chimney of the crematorium.

Unlike Treblinka or Auschwitz-Birkenau, the camp at Majdanek was built near the major urban center that would supply its victims. It was not hidden in the countryside. It is easy to imagine the smoke from the crematorium drifting into the heart of downtown Lublin. Likewise, it is hard to believe that the people of Lublin did not know what was happening at the camp. Majdanek was first established as a slave labor camp in 1940, but its gas chambers began operating in November 1942. Approximately 200,000 people were killed at Majdanek, either in the gas chamber, by shootings, or because of overwork, disease, and malnutrition. In one day alone, 3 November 1943, 18,000 prisoners were shot and killed, the bodies piled high in open ditches near the crematorium. Over 800,000 shoes were found at Majdanek when it was liberated in July 1944 by the advancing Russian army. This was the first of the camps to be liberated, the first to be seen by the Allied forces and the Western media. Unlike the camps further west that were liberated later, Majdanek was not destroyed by the retreating German forces. Although many of the wooden barracks have deteriorated through natural decay, the camp as a whole remains today as it did in 1944, relatively intact.[1]

I stand in the small open courtyard a few dozen yards beyond the entrance gate. On this spot the selections of arriving prisoners were made—who would live and work in the camp, who would be killed immediately. To my right is the gas chamber. On my left is a row of barracks, used as storerooms and work areas when the camp was in operation. These unheated and dimly lit barracks now house museum exhibits. Beyond the first row of barracks is the main camp, divided into several sections. Each section consists of two rows of barracks facing a wide open parade ground. I enter the gate and walk through the parade ground and on to the road leading to the crematorium and the site of the November 1943 mass shooting. The camp is virtually empty of visitors. As in Warsaw the day before, there is a light rain and mist, and the autumn air is cold, a harbinger of winter.

The Majdanek camp is too beautiful—the green grass of the parade ground suggests a college campus, not a site of slave labor and mass executions. Can we stand here in this lush grassy meadow and imagine the mud, the dirt, the smell—the unrelenting gray horror of the thousands of prisoners in their ill-fitting striped suits standing at roll calls? Can we imagine the perpetually gray sky, filled with smoke from the crematorium just down the road? Perhaps it would be better to see the camp in the middle of winter when one is not overwhelmed by the color of the green grass. As in the Warsaw cemetery, Nature again prevents me from seeing, understanding, and feeling the true dimensions of the remnants of the evil that confronts me.

The experience of these two places raises questions for me about the healing power of Nature—complex questions involving the ontological and normative status of Nature in its relationship to human activity, and further questions about the Nature of Nature's activity. Can the study of Nature and natural processes teach us anything about the evil of human genocide? Can the study of genocide teach us anything about the human-induced destruction of the natural world, sometimes called the process of "ecocide?" This is not a subject that will permit facile comparisons and analogies. We study the Holocaust and the environmental crisis that currently surrounds us from different perspectives—with different attitudes and purposes. Yet the comparison may be helpful. The idea of domination can be used to link together an analysis of these two evils, and can point us in the direction of developing a harmonious relationship with both the natural world and our fellow human beings.

II

Let me begin by emphasizing the importance of my visit to the actual sites described above. This essay contains more than a philosophical argument—the ideas set forth in these pages could not have been developed by me through the typically philosophical method of argument, analysis, example, and rebuttal. The lived experience of these places not only colors my ideas but to some extent completely informs them. Indeed, the essay may be merely a written expression of my attempt to come to terms with the physical experience of these places, and to place these experiences of Holocaust sites into the context of my philosophical thoughts about the meaning of the environmental crisis and the practice of human domination.

Why connect these two areas of inquiry? Why think about the environmental crisis and the Holocaust in terms of one another? Is there a meaningful relationship between human ideas of the natural world and the concepts of domination and genocide? The Nazis thought so. As Robert-Jan Van Pelt recounts in his historical investigation of the development of Auschwitz, the reconstruction and development of Polish farmland under scientific principles of management was one of the major goals of German settlement in the conquered lands east of Germany. Quoting from a contemporary record, Van Pelt describes a trip through Poland in 1940 undertaken by Heinrich Himmler, the Reichskommissar for the resettlement of the German people. Himmler and his personal friend Henns Johst stand in a Polish field, holding the soil in their hands, and dream of the great agricultural and architectural projects to come: the re-creation of German farms and villages, the replanting of trees, shrubs, and hedgerows to protect the crops, and even the alteration of the climate by increasing dew and the formation of clouds.[2] As part of this plan, of course, there would have to be an "ethnic cleansing" of the region—the Poles, both Gentile and Jewish, would have to be moved elsewhere or otherwise eliminated so that a German agricultural utopia could be developed. Thus we see that the control of Nature—the management of agriculture so as to affect even the climate—was part of the Nazi plan. The domination of Nature and humanity are clearly linked.

The goal of the domination of Nature remains with us, in the Western world, even today. As I have argued elsewhere, the primary goal of the Enlightenment project of the scientific understanding of the natural world is to control, manipulate, and modify natural processes for the increased satisfaction of human interests.[3] Humans want to live in a world that is comfortable—or at least, a world that is not hostile to human happiness and survival. Thus the purpose of science and technology is to comprehend, predict, control, and modify the physical world in which we are embedded. This purpose is easy to understand when we view technological and industrial projects that use Nature as a resource for economic development—but the irony is that the same purpose, human control, motivates much of environmentalist policy and practice.

Consider briefly, as an example, the arguments of two writers on the theory of environmental policy: Martin Krieger's call for artificial wildernesses that will be pleasing to human visitors, and Chris Maser's plans for re-designing forests on the model of sustainable agriculture.[4] Maser is an environmentalist and Krieger is not; yet their views on environmental policy are strikingly similar. Maser is considered a spokesperson and leader of enlightened environmental forestry practices, but his goal is to manage forests in such a way as to maximize the wide variety of human interests in forest development: sustainable supplies of timber, human recreation, and spiritual and aesthetic satisfaction. Krieger is a public policy analyst interested in the promotion of social justice. His goal is to develop an environmental policy consistent with the maximization of human economic, social, and political benefits. Thus he argues that education and advertising can re-order public priorities, so that the environments that people want and use will be those available at the lowest cost. Natural environments need not be preserved if artificial ones can produce more human happiness at a lower cost.

What ties together views such as Krieger's and Maser's is their thoroughgoing anthropocentrism—human interests, satisfaction, goods, and happiness are the central goals of public policy and human action. This anthropocentrism is, of course, not surprising. Since the Enlightenment, at least, human concerns—rather than the interests of God—have been the central focus of almost all human activities, projects, and social movements. The institutions of human civilization believe that it is their mission to improve the lives of human beings. Although methods may differ, and the class of people that is the primary object of this concern may differ, the central anthropocentric focus is consistent regardless of ideology or social position or political power—humanity is in the business of creating and maximizing human good.

Anthropocentrism as a worldview quite easily leads to the practices of domination, even when the domination is not articulated. In the formation of environmental policy, Nature is seen as a nonhuman "other" to be controlled, manipulated, modified, or destroyed in the pursuit of human good. As a nonhuman other, Nature can be understood as merely a resource for the development of human interests; as a nonhuman other, Nature has no valid interests or good of its own. Even the practice of ecological restoration, in which degraded ecosystems are restored to a semblance of their original states, is permeated with this anthropocentric ideology. Natural ecosystems that have been harmed by human activity are restored to a state that is more pleasing to the current human population. A marsh that had been landfilled is re-flooded to restore wetland acreage; strip-mined hills are replanted to create flowering meadows; acres of farmland are subjected to a controlled burn and replanting with wildflowers and shrubs to recreate the oak savanna of pre-European America. We humans thus achieve two simultaneous goals: we relieve our guilt for the earlier destruction of natural systems, and we demonstrate our power—the power of science and technology—over the natural world.[5]

But the domination of nonhuman Nature is not the only result of an anthropocentric worldview—the ideology of anthropocentric domination may also extend to the oppression of other human beings, conceived as a philosophical "other," as nonhuman or as subhuman. Or as C. S. Lewis wrote fifty years ago, "what we call man's power over Nature turns out to be a power exercised by some men over other men with Nature as its instrument." The reason that this exercise of power is justifiable is that the subordinate people are not considered human beings: "they are not men at all; they are artefacts."[6] Anthropcentrism does not convert automatically into a thoroughgoing humanism, wherein all humans are treated as equally worthwhile. As we have seen historically, the idea of human slavery has been justified from the time of the ancient Greeks onward by designating the slave class as less than human. In this century, the evaluation of other people as subhuman finds its clearest expression in the Nazi propaganda concerning the Jews, but we find its echoes in the ethnic civil war in the former Yugoslavia and in the continuing hatred of extreme right-wing Israelis for the Palestinians. From the starting point of anthropocentrism, domination and oppression are easily justified. The oppressed class—be it a specific race or religious group, or even animals or natural entities—is simply denied admittance to the elite center of value-laden beings.[7] From within anthropocentrism, only humans have value and only human interests and goods need to be pursued. But who or what counts as a human is a question that cannot be answered from within anthropocentrism—and the answer to this question will determine the extent of the practice of domination.

Thus the ideas of anthropocentrism and domination tie together a study of the Holocaust and the current environmental crisis. Genocide and ecocide are only possible when we conceive of our victims as less than human, as outside the primary circle of value.

III

The resurgence of trees in the Warsaw cemetery and the lush green grass of the meadow at Majdanek serve as a catalyst for rethinking the relationships among Nature, humanity, and the practice of domination. In these places, one can only describe the processes of Nature as a kind of healing, a soothing of the wounds wrought by the evil of the Holocaust. Does Nature make everything better? Can we say that dominated and oppressed entities are saved—redeemed—by the ordinary processes of the natural world? Does Nature have this power?

And if it does, what are the implications for the way in which humanity acts in relationship to the natural world?

First, we should note a possible objection to this entire line of analysis. One might argue that in thinking of Nature as having a redeeming power over human evils, we are, in part, treating Nature as if it possessed a kind of intentional activity. But Nature is not a rational subject. Nature makes no decisions, rational or otherwise. If the lush vegetation hides the horrors of Majdanek this is not the result of any natural plan, merely the effects of natural processes in their normal operations. According to this objection, we should be wary of anthropomorphizing natural processes, of being misled by metaphor and analogy.

This objection serves as an important warning to the analysis that follows. Nature has no intentions—and no other thoughts, desires, wants, or needs. Nevertheless, we can consider Nature to be analogous to a human subject. Human activities can benefit or harm natural processes in ways similar to the benefits and harms inflicted on other humans, on human institutions, and on nonhuman living beings. Moreover, Nature does act in predictable ways similar to a thinking being. As Colin Duncan has claimed, "While Nature is certainly not a person . . . it does have some of the attributes of a Hegelian subject. It can be both victim-like and agent-like."[8] Most important, we can consider Nature as the subject of an ongoing history that can be interfered with or destroyed by human action. From the perspective of normative value theory, Nature develops in ways similar to human subjects—the continuous processes of Nature produce good and bad consequences for itself and for other entities. Morally and axiologically, then, Nature can be considered to be equivalent to a subject. Without anthropomorphizing Nature—without attributing to it the emotions, feelings, and rational will of human subjects—we can understand that it is not merely a passive object to be manipulated and used by humanity.[9]

Nature, in fact, acts upon human beings, human institutions, and the products of human culture in powerful ways. So-called natural disasters, such as earthquakes and floods, are the prime examples of events in which natural forces impact on humanity. But ordinary weather, small changes in climate, and even the rotation of the earth are also activities of Nature—natural processes—that affect human life. Elsewhere I have categorized this type of activity as Nature's imperialism over humanity, for it has a parallel structure to the basic kind of human imperialism over other humans, as well as to the human imperialism over Nature. Imperialism is a form of domination, in which one entity uses, takes advantage of, controls, or otherwise exerts force over another. If we consider Nature as both a possible subject and object of imperialism, then we can think of Nature as exerting its power—attempting to dominate—humanity, just as we can think of humanity attempting to dominate Nature.[10]

But my experiences in the Warsaw cemetery and at Majdanek suggest that Nature's domination in these places is benign. It appears to heal the scars of human atrocities. Nature here does not exert the oppression of an imperialist. Nature provides the balm to restore the health and goodness of a world wounded by human evil. Nature's domination—its resurgence in these areas of human atrocities—serves as the corrective to the effects of human domination, in this case to the oppression and genocide of Eastern European Jewry. Is this an appropriate way to interpret the experiences of these places?

Perhaps not. One objection to viewing Nature as a benign healer of human-induced wounds is that such a view of Nature is yet another expression of an anthropocentric worldview. Rather than use Nature as a physical resource for economic purposes, we are here using Nature as an emotional resource, to make us feel better about the horrors of human destruction.[11] We are blinded to the fact that natural processes develop independent from human projects; Nature follows its own logic. The desire to see Nature as a healer demonstrates how pervasive is the anthropoecentric perspective. We humans seem incapable of viewing the natural world on its own terms, free of the categories and purposes of human life and human institutions.

Even more important, the question arises whether or not Nature can heal these wounds of human oppression. Consider the reverse process, the human attempt to heal the wounds of Nature. We often tend to clean up natural areas polluted or damaged by human activity, such as the Alaskan coast harmed by the *Exxon Valdez* oil spill. But we also attempt to improve natural areas dramatically altered by natural events, such as a forest damaged by a massive brush fire, or a beach suffering severe natural erosion. In most of these kinds of cases, human science and technology is capable of making a significant change in the appearance and processes of the natural area. Forests can be replanted, oil is removed from the surface of bays and estuaries, sand and dune vegetation replenish a beach. But are these activities the healing of Nature? Has human activity—science and technology— restored Nature to a healthy state?

No. When humans modify a natural area they create an artifact, a product of human labor and human design.[12] This restored natural area may resemble a wild and unmodified natural system, but it is, in actuality, a product of human thought, the result of human desires and interests. All humanly created artifacts are manifestations of human interests—from computer screens to rice pudding. An ecosystem restored by human activity may appear to be in a different category—it may appear to be an autonomous living system uncontrolled by human thought—but it nonetheless exhibits characteristics of human design and intentionality: it is created to meet human interests, to satisfy human desires, and to maximize human good.

Consider again my examples of human attempts to heal damaged natural areas. A forest is replanted to correct the damage of a fire because humans want the benefits of the forest—whether these be timber, a habitat for wildlife, or protection of a watershed. The replanting of the forest by humans is different from a natural re-growth of the forest vegetation, which would take much longer. The forest is replanted because humans want the beneficial results of the mature forest in a shorter time. Similarly, the eroded beach is replenished—with sand pumped from the ocean floor several miles offshore—because the human community does not want to maintain the natural status of the beach. The eroded beach threatens oceanfront homes and recreational beaches. Humanity prefers to restore the human benefits of a fully protected beach. The restored beach will resemble the original, but it will be the product of human technology, a humanly designed artifact for the promotion of human interests.

After these actions of human restoration and modification, what emerges is a Nature with a different character than the original. This is an ontological difference, a difference in the essential qualities of the restored area. A beach that is replenished by human technology possesses a different essence than a beach created by natural forces such as wind and tides. A savanna replanted from wildflower seeds and weeds collected by human hands has a different essence than grassland that develops on its own. The source of these new areas is different—man-made, technological, artificial. The restored Nature is not really Nature at all.

A Nature healed by human action is thus not Nature. As an artifact, it is designed to meet human purposes and needs—perhaps even the need for areas that look like a pristine, untouched Nature. In using our scientific and technological knowledge to restore natural areas, we actually practice another form of domination. We use our power to mold the natural world into a shape that is more amenable to our desires. We oppress the natural processes that function independent of human power; we prevent the autonomous development of the natural world. To believe that we heal or restore the natural world by the exercise of our technological power is, at best, a self-deception and, at worst, a rationalization for the continued degradation of Nature—for if we can heal the damage we inflict we will face no limits to our activities.

This conclusion has serious implications for the idea that Nature can repair human destruction, that Nature can somehow heal the evil that humans perpetuate on the earth. Just as a restored human landscape has a different causal history than the original natural system, the reemergence of Nature in a place of human genocide and destruction is based on a series of human events that cannot be erased. The natural vegetation that covers the mass grave in the Warsaw cemetery is not the same as the vegetation that would have grown there if the mass grave had never been dug. The grass and trees in the cemetery have a different cause, a different history, that is inextricably linked to the history of the Holocaust. The grassy field in the Majdanek parade ground does not cover and heal the mud and desolation of the death camp—it rather grows from the dirt and ashes of the site's victims. For anyone who has an understanding of the Holocaust, of the innumerable evils heaped upon an oppressed people by the Nazi regime, the richness of Nature cannot obliterate nor heal the horror.

IV

What we see in the Warsaw cemetery and the Majdanek death camp is another example of Nature's imperialism over humanity—the mirror image of the human destruction of the natural environment. Nature here acts—without an intention or design—to erase the remnants of human evil. To speak in metaphor, Nature imposes its vision of the world on its human interpreters. But Nature's vision is not our vision, and in this case it does not express the essence of the places we experience. Nature's restoration of a site of human destruction alters the character of the site, just as the human restoration of a degraded ecosystem turns a natural area into an artifact. Although the beauty of the trees in the cemetery cannot be denied, the meaning and value of the cemetery lies not in the trees but in the historical significance of the Nazi plan to kill the Jews of Eastern Europe.

Nature's reemergence at these Holocaust sites is a form of domination: the domination of meaning. Nature slowly exerts its power over the free development of human ideas, human history, and human memory. Now it may seem strange to think of the healing power of Nature—the healing power of anything—as a form of domination. But Primo Levi describes his liberation from Auschwitz in terms that suggest this relationship.[13] He recounts the series of baths that he and the other prisoners were given by the Allies: "it was easy to perceive behind the concrete and literal aspect a great symbolic shadow, the unconscious desire of the new authorities, who absorbed us in turn within their own sphere, to strip us of the vestiges of our former life, to make of us new men consistent with their own models, to impose their brand upon us."

But Levi also compares these baths of liberation with the "devilish-sacral" or "black-mass" bath given by the Nazis as he entered the universe of the concentration camps. All of these baths served as symbols of domination—the molding of human beings into artifacts appropriate for their current situations. The cleansing of liberation is thus comparable to the oppression of imprisonment, for both actions deny the autonomy of the free human subject. Healing thus can be an expression of domination, if it modifies or destroys the meaning and the freedom of the original entity.[14]

To understand the multiplicity of the forms of domination, however, is the first step toward developing a comprehensive ethic for evaluating human activity in relationship to both the natural environment and the human community. We must resist the practice of human domination in all of its forms. We must act so as to preserve the free and autonomous development of human individuals, communities, and natural systems. We must understand the moral limits of our power to control Nature and our fellow human beings.

And so I am reminded of the last verse of the kaddish, the prayer that closes all Jewish services, and also serves as the prayer of mourning for the dead. This verse is a call for the healing power of peace. *Osay shalom bimromov hoo ya-ahsay shalom, olaynoo v'al kol yisroayl.* "May He who establishes peace in the heavens, grant peace unto us and unto all Israel." In viewing the Warsaw cemetery and the Majdanek death camp, I was moved by the hope that Nature could be the agent who establishes peace. But Nature alone cannot accomplish this. If there is a God, He works through human decisions. Only humans can understand the meaning and history of evil. Only humans who understand the need to control our power can halt the practice of domination, can halt the destruction of people and the natural environment. It is only through human actions that peace can be restored to our planet and our civilization.

Notes

1. For a general discussion of Majdanek and the overall history of the Holocaust, see Leni Yahil, *The Holocaust: The Fate of European Jewry*, trans. by Ina Friedman and Haya Galai (New York: Oxford University Press, 1990), especially 362–63; Ronnie S. Landau, *The Nazi Holocaust* (Chicago: Ivan R. Dee, 1994); and Martin Gilbert, *The Holocaust: A History of the Jews of Europe during the Second World War* (New York: Henry Holt, 1985). The death statistics cited in these recent works differ by an order of magnitude from Dawidowicz's classic work, which claims that 1.3 million Jews died at Majdanek (Lucy S. Dawidowicz, *The War against the Jews, 1933–1945* [New York, Holt, Rinehart, and Winston, 1975], 149). Gilbert reports that Hitler was enraged that the German SS forces did not destroy the camp before the Russian advance (711).

2. Robert-Jan Van Pelt, "A Site in Search of a Mission," in *Anatomy of the Auschwitz Death Camp*, ed. Yisrael Gutman and Michael Berenbaum (Bloomington: Indiana University Press, 1994), 101–103.

3. See Eric Katz, "The Call of the Wild: The Struggle against Domination and the 'Technological Fix' of Nature," *Environmental Ethics* 14 (1992): 265–73; "Artefacts and Functions," *Environmental Values* 2 (1993): 223–32; and "Imperialism and Environmentalism," *Social Theory and Practice* 21:2 (summer 1995): 271–85.

4. See Martin Krieger, "What's Wrong with Plastic Trees?" *Science* 179 (1973): 446–55; and Chris Maser, *The Redesigned Forest* (San Pedro: R&E Miles, 1988).

5. See Eric Katz, "The Big Lie: Human Restoration of Nature," *Research in Philosophy and Technology* 12 (1992): 231–41, and "Restoration and Redesign: The Ethical Significance of Human Intervention in Nature," *Restoration and Management Notes* 9, no. 2 (1991): 90–96.

6. C. S. Lewis, "The Abolition of Man," reprinted in *Philosophy and Technology: Readings in the Philosophical Problems of Technology*, ed. Carl Mitcham and Robert Mackey (New York: Free Press, 1983), 143-50, especially. 143 and 146. Lewis's *The Abolition of Man* was originally published in 1947.

7. Thus the power of Peter Singer's argument that animal liberation is necessary to correct speciesism, a prejudice akin to racism or sexism (Peter Singer, *Animal Liberation: A New Ethics for Our Treatment of Animals* [New York: Avon, 1975], 1–23).

8. Colin A. M. Duncan, "On Identifying a Sound Environmental Ethic in History: Prolegomena to Any Future Environmental History," *Environmental*

History Review 15:2 (1991): 8. See also Katz, "The Call of the Wild," and "Imperialism and Environmentalism."

9. See Katz, "Imperialism and Environmentalism," 274. Holmes Rolston III presents a sustained account of the idea of Nature as the subject of an ongoing history (Rolston, *Environmental Ethics: Duties to and Values in the Natural World* [Philadelphia: Temple University Press, 1988], esp. pp. 342–54).

10. See Katz, "Imperialism and Environmentalism," esp. 273–74.

11. I am indebted to Avner de-Shalit for bringing this argument to my attention.

12. The argument in this section is based on Katz, "The Big Lie," "Call of the Wild," and "Artefacts and Functions."

13. Primo Levi, *The Reawakening*, trans. Stuart Woolf (New York: Collier Books, 1987), 8.

14. Although it may appear paradoxical to think of the act of healing as a form of domination, consider the long standing issue of paternalism in the field of medical ethics. The use of medical procedures against the wishes of a fully rational patient is a violation of individual autonomy, even when these medical procedures are clearly in the best interests (i.e., the health) of the patient.

The Takings Clause and the Meanings of Land

Zev Trachtenberg

I

The Fifth Amendment to the United States Constitution states that "private property [shall not] be taken for public use, without just compensation." In recent years—and reaching a crescendo after the 1994 elections—a chorus of voices has insisted that the "takings clause," as this brief phrase is called, renders unconstitutional much of the government's effort to protect the environment by regulating private land use. Federal rules governing wetlands have provided a salient target for this attack; aggrieved property owners claim that requiring permits for any construction or filling effectively denies them control over their land.[1] With these rules, they hold, the government has taken something of value from them, and they are owed compensation.

Compensating owners for their lost value is not simply the constitutional thing to do, on this argument—it is a matter of justice. Thus, the case against wetlands regulations asserts that they deprive property owners of the benefits they had expected from the free use of their property—say, building a home on it, or using it to grow crops. Perhaps these regulations have a worthy purpose: maintaining the benefits wetlands provide for the public. But, the argument goes, the government has forced property owners to subsidize, through their foregone personal benefits, the maintenance of these public benefits. This seems manifestly unjust: the owners have done nothing to deserve the imposition of what is really a special tax, levied only on them. Rather, if they must, in the public interest, forgo the benefits from their land, they deserve to be compensated for their loss. As a matter of justice, if the public receives a benefit from private parties, the public ought to pay for it. If the public regards having pristine wetlands as a benefit, it ought to purchase them from their owners. Indeed, this is the rationale behind the constitutional mandate for compensation.[2]

But wetlands policy is just one front in a broad conflict over property rights that also encompasses disputes over zoning, historic preservation, and land use in general.[3] I am concerned with the general issue of the clash between property rights and environmental protection, and will treat wetlands policy as a case in point, rather than a topic for specific investigation. How should we think of this clash as a matter of justice? As we have just seen, advocates of property rights can argue that, insofar as they constitute uncompensated takings, environmental regulations are inherently unjust. But environmentalists can argue in response that such regulations are perfectly just. Indeed, they hold, it would be unjust to allow individuals to use land in ways that harm the public as a whole; filling wetlands, for example, destroys their ability to purify water, to control floods, and to provide habitat for wildlife.[4] It is unjust, that is, for the public to suffer the losses that result from environmental degradation; justice demands that the government regulate the activities of individuals that pose serious threats to others' safety and health.[5] The injustice would be compounded by the requirement that compensation be paid to those whose activities are regulated; compensation would be a kind of blackmail, paid to avert a threatened harm.

How then do the scales of justice fall in the conflict between property rights and environmental protection? To answer this question I will examine environmental regulation as a matter of distributive justice. Distributive justice has to do with the rightful distribution of entitlements between private individuals; the great question of distributive justice is whether (and how) inequality between people can be justified.[6] I shall, however, consider a different domain of distribution: the distribution of entitlements between private individuals and the public.

To explain what I have in mind here, let me make use of the legal notion that ownership consists of a number of "sticks" in a "bundle of rights."[7] These individual rights are distinguishable from each other, and perhaps from title to the land. For example, as it happens I own the surface of my land, but someone else owns the subsurface mineral rights. Further, I do not own the right to erect a skyscraper on my land, since my neighborhood is zoned for low-rise buildings. Considerations of distributive justice can apply to particular "sticks" as much as to "bundles" as wholes. I do not question the justice of an oil company owning the minerals beneath my house since it purchased this right from a previous owner, and the diminished value of my land was reflected in the price I paid to buy it. The question of zoning is more contentious; let me say simply that I hold it as just that my town has asserted its control over the right to build above a certain height in my neighborhood, since I think that valuable ends are thus served.[8]

We can construe the question of environmental regulation, then, as a question about the proper distribution of certain land use rights. Ought the right to fill a wetland, to pursue our example, inhere in the title holder, or ought it rest with the public? To assert that regulations are unjust is to affirm the former view: compensation is required because the regulation has effected a taking of something that justly belongs to the landowner. And to affirm the latter view is to assert the justice of regulations: the regulation puts into effect the public's just control over the right in question, by allowing the public to assign the right, suitably regulated, to the landowner through the permitting process.

But how are we to determine the just distribution of those rights over land that are contested by property owners and environmentalists? Which rights would a just distribution place squarely in the hands of title-holders, and which would it place under public control? Michael Walzer has famously suggested that questions of distributive justice must be settled by examining the social meanings of the goods at issue. In his book *Spheres of Justice*, Walzer argues that attention to a given society's practices regarding a given social good reveals the meaning of that good shared by society's members. The criteria for just distributions of different goods flow out of each good's distinct meaning. For example, he interprets the way health care is organized in the United States as showing that this society assigns health care to a certain "sphere" of life, the sphere of basic human welfare, which is to be distributed according to need.[9] In this sphere, distributive justice does not mandate, for example, an equal distribution of health care services; it is entirely just that people who are more seriously ill receive more attention from doctors than their healthier neighbors.

In this chapter I shall take up Walzer's suggestion, and look to the realm of meanings in order to think about where justice lies in the matter of environmental regulation. But a problem immediately arises with Walzer's approach. For resolving the issue at hand is not a simple matter of determining the distributive criterion that flows out of the meaning of land—as if a good as fundamental as land had a single meaning, agreed on by all members of society. Of course this is not the case. Land has a variety of meanings, with conflicting imports. In a moment I will survey three broad meanings land has in contemporary American society. The first two frame land as individual property—either as a marketable commodity, or as holding invested with personal significance. These two meanings I take to be quite manifest, and to be now enjoying political prominence. They are the meanings that support the assignment of a great range of land use rights to individual titleholders. The third meaning associates land with community; it assigns these rights to the public. Part of my purpose here is to reaffirm the weight of this latter meaning—to insist that it be acknowledged along with the first two.

But before proceeding we must observe that the multiplicity of meanings of land suggests that my survey will not reveal a single authoritative distributive criterion that can answer my question. Indeed, that important social goods have a plurality of meanings has been a point emphasized by Walzer's critics. Ronald Dworkin argues that the most important goods are the subject of intense debate, hence that there simply are no shared understandings as to their meanings.[10] Walzer responds that the debates Dworkin cites can reflect competing interpretations of the underlying meaning, or debates on how to distinguish between "overlapping or entangled goods."[11] But as Georgia Warnke observes, Walzer's response does not rule out the possibility that the underlying meaning is itself multidimensional.[12]

How then should goods with multiple meanings be distributed? What is needed is some scheme for resolving these competing meanings into one. We must concede that if we are to follow Walzer in looking to social meanings to provide distributive criteria, when considering multiple meanings we are obliged to produce a single result: the takings issue is a practical matter in which distributive decisions must be made. Further, we might well think, the result we produce must hold good over all takings cases: that is, that this distributive criterion (and no other) is the right one, and is the only one that ought to be applied. For one of our most fundamental political ideals is that of the rule of law, which incorporates the belief that judges ought to decide cases consistently, on the basis of impartial principles.[13] This ideal is particularly important in the area of property: society could hardly go on if people were uncertain of the extent of their rights over their holdings.[14] Uncertainty would be a natural result if, for example, a judge in a takings case could decide arbitrarily which meaning of land to honor, hence which distribution of rights to enforce. To avoid arbitrariness in the administration of the property regime, therefore, it seems necessary to settle on one meaning as definitive.[15]

One way to achieve consistency of decisions given multiple meanings would be to prioritize, to rank the meanings according to some standard.[16] We might be committed to some moral principle that we can apply to each meaning in order to judge which is the "most" moral, for example.[17] However, it is quite uncertain that the meanings of land can be ranked in this fashion. I assume (and hope to demonstrate) that each of the meanings of land I shall survey has a strong claim on us, yet none is so strong that we would be comfortable giving it absolute priority over the others. Certainly none is more "authentic" than the others. Thus, it strikes me that they are like what Brian Barry calls incommensurable values.[18] That is, there is no single standard that reveals an order, or even a fixed scheme of weighting, among these meanings. There is no "measuring stick" we can apply to each meaning, in order to determine which is the best, across all cases.

As we think about the meanings of land, therefore, we are in the position David Miller describes: "we find we have to arbitrate between several competing values, none of which we feel inclined, on reflection, to abandon."[19] Hence, if we were forced to choose one meaning as definitive, we would be forced to abandon the others; that is, we would be forced to sacrifice some values for the sake of one that is held to be preeminent. Situations of this sort are tragic situations.[20] Indeed, it appears, the takings issue has a tragic quality, at two levels. At the level of theory, if our commitment to the rule of law forces us to choose one meaning to honor in legal decisions, we will dishonor meanings that deserve to be attended. At the level of individual decisions, even if these are made "intuitionistically," that is, by choosing the meaning that on balance seems best to fit the specific circumstances, again important values associated with the other meanings are lost.[21]

The existence of multiple meanings of land thus profoundly complicates our attempt to understand where justice lies in the conflict between property rights and environmental protection. Let us proceed, then, to examine the meanings I have in mind. We will then return to consider the tragic quality of the takings problem, and the challenge it poses to the rule of law.

II

Let us begin with what is perhaps the most available meaning, suggested by the practice that standardly governs the relation between the public and private rights over land. This is the practice of eminent domain, that is, paying just compensation when the government takes land for a public purpose. The accepted standard for just compensation is fair market value.[22] If the government exercises its power of eminent domain, say by condemning my land to build a road, it must pay me what my land is worth, defined as the price on which a willing seller and a willing buyer would agree. What, then, does our compensation practice suggest is the meaning of land? The concept of fair market value, obviously, applies in virtue of land's status as marketable property. Thus, the practice of compensation points to the meaning of land associated with the institution of the market: land is a commodity. Since the practice of buying and selling land is essential to our society, it goes almost without saying that the notion of property is at the core of our shared understanding of land.

Let us interpret this meaning further. In the context of the market, the meaning of land (like that of any other commodity) is that it is a bearer of financial value determined by free exchange. Margaret Jane Radin refers to this form of property as "fungible" property; she contrasts fungible with "personal" property, to which I will return below.[23] The value of fungible property is not connected to any significance its specific characteristics might have for its owner; these are valuable only insofar as they have value to others, to whom the property can be sold. Thus, one's various bits of fungible property are interchangeable one with another. They all have the same meaning, as abstract as money's: the mere possibility of exchange. Fair market value, in this light, makes for perfectly just compensation. The money fully substitutes for the taken land, since the price one receives has precisely the same value as the land itself, understood as fungible property.

The idea that what gives land meaning is its ability to bear financial value is rooted in Locke, perhaps the most influential source of Americans' shared understanding of landownership.[24] His treatment of land is complex: its layers are revealed by considering the meaning Locke attributes to labor. In the opening paragraphs of chapter 5 of the *Second Treatise*, land represents the opportunity for survival: God gave the earth and its bounty to mankind so that mankind would not simply perish. But survival is not a free gift: Locke insists that God really gave the earth "to the use of the industrious and rational."[25] Thus, individuals must labor to gain their needs from nature, and those who labor demonstrate their rationality. But their reward is more than bare survival: the things on which they labor become their property as a matter of natural right. Although the simplest examples of property are fruits or animals taken from nature, Locke quickly asserts that "the chief matter of property" is "the earth itself" (§ 32). Land becomes property through the labor that renders it productive; indeed, Locke holds that without labor land is practically valueless (§ 43). Thus, to begin with, land means the chance rationally to pursue survival through labor.

But this rather personal meaning is only a stepping stone for Locke. While the rational inhabitant of Locke's state of nature recognizes the necessity to labor, he (Locke is speaking of males) also recognizes that there are natural law limits to the amount of property he may amass. In particular, he may not allow anything in his possession to spoil; one may rightfully possess only what one can consume (§ 31).[26] But Locke presumes that the rational individual will wish to have as much property as possible, without violating natural law. The way out of this impasse is offered by the possibility of exchanging perishable for imperishable goods: in Locke's example, plums that go bad in a week for nuts that go bad in a year, or even better, for shiny pebbles that never go bad at all (§ 46). Someone "might heap up as much of these durable things as he pleased; the exceeding of the bounds of his just property not lying in the largeness of his possession, but the perishing of anything uselessly in it" (§ 46).

"And thus came in the use of money, some lasting thing that men might keep without spoiling, and that by mutual consent men would take in exchange for the truly useful, but perishable supports of life" (§ 47). But note that, for Locke, it is labor that gives the useful things value. An apple is valuable to me only if I have labored to get it from the tree so that I can eat it; on the tree it has no value at all. If I let the apple spoil, I will have lost that value, the value of my labor. If I trade the apple for a shiny pebble, the pebble holds that value—so that it can be traded whenever I please for something else. The function of money, then, is to store the value created by labor. Without money it is irrational to labor beyond the point of providing for immediate consumption. The institution of money allows for the full play of rationality, since its product, the value created by labor, can be maintained into the future, when it can be exchanged for some other good.

We are now in a position to see the fuller meaning of land in Locke's account. Given the institution of money, the meaning of land is no longer restricted to mere survival; land is no longer simply a resource for producing the supports of one's own life. Now land provides the opportunity for the indefinite accumulation of value. I use my land to create value by growing crops; I store the value by exchanging my crops for money. Since I may now grow more than I need, I can exchange the surplus for more land, and the labor of employees, and for capital equipment—all in order to create more value, which again can be stored and reinvested. As C. B. MacPherson argues, Locke valorizes the making of profitable exchanges as the clearest evidence of rationality, hence, as the fulfillment of God's intentions for human beings.[27] The meaning of land emerges from this broad conception: for Locke, land means the possibility of producing value for use in exchange. That is, for Locke, ultimately land means fungible property—it is to be understood as a commodity.

Now it seems clear that the practice of eminent domain in the United States is based on this Lockean understanding of land as fungible property. As Radin puts it, "in assuming that compensation is an appropriate corrective measure, that it can be 'just' or make owners whole, the current idea of eminent domain assumes that all property is fungible."[28] In light of this meaning—land as commodity—let us return to the question at hand: what criterion does it suggest for determining the just distribution between private owners and the public of rights over land?

On the one hand, within the Lockean tradition, the understanding of land as a commodity supports the view that, as a starting point, rights over land vest in its owners. This view is based on a (frankly hypothetical) historical account, which looks back to a moment of original acquisition when items in nature first become private property. In Locke's version, land becomes property through an original appropriator's labor, which grants full ownership rights, subject only to natural law strictures such as that against spoilage.[29] "There is then," in Richard Epstein's words, "a unitary conception of ownership which flows comfortably from the doctrine of original acquisition."[30] Specifically, the right of disposition—the right of the owner to do with the land what he or she will—is fully part of the original property right.

But, given an understanding of land as a commodity, this distribution of all rights over land to its owner need be only a starting point; the practice of eminent domain explains how the public may rightfully obtain certain land rights. For consider that the criterion that distributes commodities among participants in a market (the institution for distributing commodities) is the ability to pay. Thus, when the government chooses to exercise its eminent domain power, it acts out of the confidence that it can afford, in effect, to purchase the rights it seeks to control. To be sure, eminent domain is the power to compel a sale—a power not (strictly) available in ordinary market transactions. However, it is broadly acknowledged that where the benefit to the public is sufficient, compulsion of this sort is justified—given that the original owner is compensated with the fair market value of the rights he or she has lost. Thus, the meaning of land as fungible property starts with an original distribution of rights to individual owners, but, in light of the conceptual ease of compensating the loss of a commodity, can easily account for a transfer of those rights to the public.

III

That the meaning of land as commodity, as fungible property, seems so available to us is perhaps because it is tied to the idea of the individual. As a commodity, land has an owner, and the social role of owner is one that we instantly recognize. But there is another meaning for land associated with the idea of the individual—a meaning that is in tension with its meaning as commodity. To introduce this second meaning let us recall Radin's distinction between fungible and personal property. Personal property is connected in an intimate way with human flourishing, which it supports in virtue of its specific characteristics; for example, one's home "is the scene of one's history and future, one's life and growth . . . one embodies or constitutes oneself there."[31] Land, of course, is especially capable of coming under the rubric of personal property, since a person's relation with the land he or she owns can play an especially powerful role in his or her personhood. For example, few notions are as powerful as that of the family homestead: a place that forms one's identity, where one feels connected to who one is. One's possession of such places is not, as with fungible property, an opportunity for exchange, but a condition for selfhood. Let us explore, then, land's meaning as personal property.

This second strand of meaning is, like the first, knotted to Locke's labor theory of property. As a matter of natural law, Locke holds, when a person "mixes" something that is properly his own—the labor of his body—to an item owned in common, he establishes a rightfully exclusive claim to it. That is, he makes it his property (§ 27). Now as we saw, Locke develops his view in a way that emphasizes the fungible character of the value produced by labor: ownership of land means the ability to produce and store up value in the form of money. But the act of laboring on land produces a far less abstract meaning as well. Land that has been worked takes on the meaning of the lives that worked it.

The notion that labor yields a deep personal connection to the land, and that this connection grounds a criticism of the conception of property as purely fungible, was articulated strongly by the "Southern Agrarians" in the 1930s. "The agrarian discontent in America," according to John Crowe Ransom, "is deeply grounded in the love of the tiller for the soil, which is probably . . . one of the more ineradicable human attachments."[32] This personal attachment to the land is clearly at odds with the "industrial" vision of land as strictly fungible property. Thus, Ransom elevates the yeoman farmer as an ideal figure to be deployed in criticism of the "industrial" way of life. The farmer, Ransom declares,

> identifies himself with a spot of ground, and this ground carries a good deal of meaning. . . . A man can contemplate and explore, respect and love, an object as substantial as a farm or a native province. But he cannot contemplate nor explore, respect nor love, a mere turnover, such as an assemblage of "natural resources," a pile of money, a volume of produce, a market, or a credit system. It is into precisely these intangibles that industrialism would translate the farmer's farm. It means the dehumanization of his life.[33]

This conception of the relation between farmer and land grounded the Agrarians' support for widespread landownership.[34] Paul Conkin cites the "proprietary ideal" as the core value of the Agrarian movement. He argues that they campaigned for "the humane concerns so well expressed in the ancient right to property."[35] That is, for the Agrarians, land meant the possibility of "the most humane of all the modes of human livelihood," a possibility realized through ownership.[36]

We should note that Ransom includes as part of the "humane" identification with the land the attitudes of respect and love. Although he does not develop this theme, he seems to oppose the strictly instrumental treatment of land by industrialism, recognizing instead something like land's intrinsic worth. When nature is not seen as a collection of resources, it can be appreciated for itself. It would no doubt be wrong to say that Ransom adopts a nonanthropocentric position—but such a position is visible from his.[37] Indeed, it is frequently argued by advocates of property rights that individual owners have greater incentive to act as stewards, protecting their land's environmental integrity. Certainly this claim points to economic self-interest: owners will not want to destroy the economic value produced by the environmental features of their land.[38] But more, it is claimed that individual owners are in the best position to appreciate these environmental values for their own sake: ownership places them in a particularly intimate relationship with "their" piece of nature. Thus, in this view, the goal of environmental protection is in fact best pursued by giving full range to personal property rights, rather than trying to distribute certain elements of control over land to an impersonal public.

But we must also recognize that this personal meaning of land has a distinctly political dimension—one that flows out of Locke's conception of property as a natural right. In Locke's account, rightful property claims are established in the state of nature, in advance of the institution of government. Thus, property is not a positive right, granted by the state. Rather, the state is created for the purpose of protecting preexisting natural rights, which Locke denotes generally with the word "property" (§ 123). The advantage of civil society is that society can collectively enforce its members' natural rights. If government itself violates natural rights, in particular by taking its citizens' property, it is rendered illegitimate and is subject to rightful overthrow (§ 222). Thus, for Locke, property is the foundation of liberal government, that is, the type of regime that respects and preserves the freedom of its citizens.[39]

As heirs of Locke, then, for us property also means liberty, in particular liberty from intrusion by the government. Jennifer Nedelsky discusses this meaning as part of what she calls the myth of property in the American constitutional tradition.[40] She notes that the legal treatment of property rights has varied widely over U.S. history, and indeed property has now come to a point where it is seen by courts as highly dependent on governmental decisions. "Nevertheless, this judicial practice does not seem as yet to have shaken the popular force of the idea of property as a limit to the legitimate power of government."[41]

As it happens, the current political debate over federal wetlands policy provides a clear example of the present force of the idea that property defines a limit to government intrusion. The Fairness to Land Owners Committee (FLOC), an organization claiming to represent nine thousand "mom and pop" landowners suffering under existing wetland regulations, offered the following statements at a 1992 congressional hearing on a proposed revision to the law:

Under the guise of protecting the environment, these bureaucrats . . . have launched the greatest war against private property rights in the history of this Nation. All over this Country, we, who own the land, have organized a backlash and a rebellion and we believe that the courts are beginning to *recognize the difference between conservation and confiscation*. . . . You, our elected representatives, have failed to protect the masses from *"a long train of abuses and usurpations"* (*The Declaration of Independence*) that possibly exceed those of the dastardly King our fore fathers overthrew in 1776. You have failed to protect the landowners from bureaucratic abuses and seizure of their land.[42]

The theoretical standpoint underlying these sentiments was expressed by Don Young, U.S. Representative from Alaska, who opined that "it is the concept of private property and respect of property rights that is one of the cornerstones of our free and democratic society."[43] The rhetoric here is perhaps overheated, but deeply instructive. For these rhetorical appeals are addressed precisely to common understandings, to meanings that have great force. The meaning of land as personal liberty, in other words, powers the sense of injustice that this rhetoric expresses, and strives to induce.

In invoking the language of the Declaration of Independence (words that echo Locke's language in the *Second Treatise* [§ 225]) the present-day defenders of property rights invoke an understanding of property that is strongly associated with Thomas Jefferson. The central place of Jefferson's ideas in American attitudes toward landownership is manifest. To explain Jefferson's conception, Eugene Hargrove cites his admiration for Saxon common law in pre-Norman England. Jefferson rejected the feudal conception that in America land belonged ultimately to the crown, and was merely held in trust by its nominal owners. Rather, Hargrove argues, "he consistently spoke of allodial rights . . . which refers to an estate held in absolute dominion without obligation to a superior."[44] In asserting that property owners held allodial rights, that is, Jefferson emphasized that ownership of land means independence and freedom.[45]

What does the understanding of land as personal property suggest ought to be the criterion for distributing rights over land between individuals and the public? Obviously, the FLOC insists that our traditions vest control over land in the title-holders. To the extent that the public is allowed to seize rights over land, it must compensate the private parties in whom the rights originally (and properly) inhered. The alternative, vesting certain rights over land in the public, allows the public's agent—the government—veto power over the title-holder's use of his or her own land: confiscation in fact, if not in name. This public control over private land runs counter to the strand of meaning that holds that ownership of land means freedom from government interference. The FLOC argues, in effect, that the just distribution of rights over land assigns them not to the public, but to the individual. Implicitly they justify this distribution by appealing to the meaning of land as liberty.

We must notice, however, that in spite of the fact that both meanings we have surveyed imply that rights over land should vest originally with owners, the second meaning is less comfortable than the first about transferring rights to the public. The difference enters with the notion of what constitutes just compensation when those rights are taken by the government. Tautologically, as we saw, where land has the meaning of fungible property, fair market value is an adequate replacement for it. But it is not at all clear that fair market value, or indeed any other price, can truly replace land that has deeply personal meaning. There is some dissonance, that is, between the meanings associated with land as fungible and as personal property.

This dissonance is exploited in an attack on eminent domain practice by the libertarian philosopher Ellen Frankel Paul. "Surely," she asks, "if the government condemns your property the just compensation language offers protection, but will you be made whole again? Certainly not if your particular piece of property held special meaning for you."[46] She cites as an example a takings case in Texas, where a number of ranchers would suffer "the uncompensable psychological loss of their land, to which they had formed deep and abiding attachments."[47] Her comments challenge the notion that compensation by fair market value is just: " 'fair market value' will seem of little comfort to you if you never wished to sell in the first place."[48] Paul's observations suggest that not only ought rights over land be distributed to owners, ideally the public ought to have a very limited ability to claim them.[49] This position renders environmental regulation highly problematic, even if it includes compensation.[50]

IV

If land means property—either fungible or personal—its meaning includes the idea that it has an owner: an individual whose relation to the land is the basis of its significance. In the current political climate in the United States, these "individualistic" meanings reverberate quite strongly, and have been used by opponents of environmental regulations to fuel their antiregulatory rhetoric. But, as of this writing, it appears that this rhetorical attack has stalled (the fate of the associated legislative agenda is less clear). For, I believe, the individualistic meanings of land have run up against another meaning that also has great power: a meaning that associates land with the idea of community. It is part of our shared understanding that land is a nexus through which the seemingly private activities of individuals can have "external" effects on others. This "communal" meaning supports the distribution of certain rights over land to the public, to guard against unwanted externalities. Some of our legal practices that recognize the communal meaning of land are the target of attack, for violating the individualistic meanings. I hope to show first that the communal meaning is as fully warranted as they are; this will demonstrate how an appeal to the communal meaning can justify environmental regulations of property. In the next section I will return to the question of the relation of the three meanings I consider, and how they might be resolved.

The communal meaning of land is at the heart of Aldo Leopold's "land ethic": he urges us to consider land as a community, of which we are also a part. "A land ethic changes the role of Homo sapiens from conqueror of the land-community to plain member and citizen of it."[51] We shall see in a moment just how Leopold understands human beings to be members of the land community. First, let us note how Leopold uses the image of land as community to highlight human beings' ethical obligations toward the land itself, conceived as an ecological system of living beings and inorganic materials. This is the lesson of his famous dictum that "a thing is right when it tends to preserve the integrity, stability, and beauty of the biotic community" (224–25), implying that the land has value of its own, apart from its economic value as property.

Leopold's dictum implicitly criticizes an understanding of land that attends only to its meaning as property. He notes that Odysseus could hang his disobedient slave-girls because they were property, and "the disposal of property was then, as now, a matter of expediency, not of right and wrong"(201). But although ethics have progressed to the point where human beings may no longer be owned, "land, like Odysseus's slave-girls, is still property. The land-relation is still strictly economic, entailing privileges but not obligations" (203). Still, but not rightfully. To the extent that we acknowledge land's meaning as community, we must recognize obligations toward it. Hence, Leopold intimates, we must give up some of the privileges that come with conceiving land simply as property. "We abuse land because we regard it as a commodity belonging to us. When we see land as a community to which we belong, we may begin to use it with love and respect" (viii).[52]

Thus there is an element in Leopold's thought that seems to encourage recognition of the intrinsic value of land.[53] It seems that to understand land as community is morally to decenter human beings: the measure of value is no longer human good. Indeed, for Leopold, the measure of the value of land is not the economic good of its owner. But, I believe, it is not right to say that Leopold adopts a strictly nonanthropocentric view. For consider why he thinks we ought to recognize obligations toward land. Why is it right, for him, to use land with love and respect? In sum, why must we attend to the communal meaning of land as fully (if not more fully) as to its meaning as property? To answer these questions we must return to the issue of how he conceives the human community with the land.

Leopold's normative imperative rests on his descriptive use of the idea of land as community. As a description, the idea serves to stress the complex connections among the elements—"soil, waters, plants, and animals" (204)—that collectively constitute the land.[54] Thus Leopold speaks of land as a "biotic mechanism," whose diverse components are bound together by the exchange of energy. "Land, then, is not merely soil; it is a fountain of energy flowing through a circuit of soils, plants, and animals" (216). The image of an energy circuit allows Leopold to emphasize the interrelatedness of the components. Thus, "when a change occurs in one part of the circuit, many other parts must adjust themselves to it" (216). Although land has always been subject to changes, and has responded to them, technology has enabled human beings "to make changes of unprecedented violence, rapidity, and scope" (217). Most frequently, he observes, the effects of such changes, the way the land responds to them, cannot be foreseen and are likely to be undesired (217–18, 205–6).

Leopold's recognition of land's interconnectedness leads to the belief that it is morally wrong for individual landowners simply to bend their land to their will like conquerors, since they might well thereby cause changes that damage themselves or others. For, as ecological knowledge has developed and disseminated, it has become more broadly realized that what seem like localized uses of the land can have very wide-scale deleterious effects.[55] With this knowledge has come the idea of "environmental services": the idea that natural areas perform many functions that support human life in previously unrecognized ways. In the example we have followed, that of wetlands, it is now understood that they help maintain water quality, and absorb excess water during floods.[56] The increasing awareness of environmental services is entering into our shared understanding of land. Hence we have begun to recognize the value of abandoning the role of conqueror of the land; we are coming to admit that when landowners act instead as members of the land-community and respect its natural processes they are less likely to do harm.[57]

Leopold's view, therefore, has not abandoned anthropocentrism.[58] The communal meaning of land provides for a human communion with intrinsic features of nature. But at bottom, the human community with the land is based on the land's transmission, through natural processes, of the effects of human actions on other human beings. Leopold's accomplishment is to bring consideration of the land into the ethical conception of the human community, so that actions affecting it are understood to be governed by familiar anthropocentric ethical values (e.g., do not harm other people). Indeed, Leopold helps us see land as the substrate of human community. This is not simply in the immediate sense that human beings ultimately rely on the land for the means of their survival.[59] More profoundly, when we recall that human community entails human connectedness, we can come to understand natural processes, mediated by the land, as another medium of human connection.

Leopold's vision of land as community has been developed by legal scholar Joseph Sax to defend the existence of "public rights" over certain kinds of land use. Sax observes that "the ecological facts of life demonstrate a powerful inextricability in the utilization of natural resources."[60] The complex web of connections between different elements in the natural world produces a wide and diverse set of results; thus an action taken on one person's land typically produces "external" effects, that is, in places beyond his or her own property. Now if the external effects cause substantial damage to a small number of others, they are likely both to recognize and take action against the responsible owner, say through a suit for nuisance. But, Sax notes, "one characteristic of external effects . . . is that they often fall quite broadly, affecting a large number of potential claimants, each in relatively small amounts."[61] Filling a wetland, resulting in a slightly decreased water quality for many, many people, would be such an action that affects a "diffuse public."

But with a diffuse public it is likely to be impossible, practically speaking, for all the affected parties to unite to initiate a private suit against the responsible owner.[62] Say then the government issues a regulation to protect the interests of the diffuse public. This sort of action is now challenged as a taking, insofar as it limits the right of the landowner to free use of his or her property. But, Sax holds, "to the extent that the courts adopt this perspective, they deny recognition of extant public rights."[63] Instead, he argues, our legal system ought to "recognize diffusely-held claims as public rights, entitled to equal consideration in legislative or judicial resolution of conflicting claims to the common resource base."[64] Thus, in Sax's view, rights over land ought not be vested in landowners as a matter of their property right; the public's rights deserve equal consideration.

But how then are conflicting claims to resources to be resolved—or, in the terms of my argument, how are rights over land to be distributed between owners and the public? For Sax, equal consideration implies that there is no a priori distribution to one side or the other. Rather, from his perspective, land's meaning as community suggests the utilitarian criterion that those rights over land that have wide-ranging effects on many people should be distributed in a way that provides for a minimization of the potential danger.[65] But insofar as it is agencies of the public—the courts or the legislature—that are charged with the responsibility of determining what distribution minimizes the possibility of damage, in effect our political practices place these rights in the hands of the public. This placement is justified by our belief that our rights must be distributed so that no one has the right to harm another, and our recognition that self-interest makes the individual an inappropriate judge of which of his or her own actions are harmful. Thus, under the "social contract" human beings are said to have created in order to enjoy the benefits of mutual cooperation, individuals are said to have assigned control over those rights that might easily lead to harm to others to the public as a whole.[66] The public, acting through its government, regulates the exercise of these rights—assigning them back to individuals through permits, or forbidding their exercise outright.

Thus we give government the role of policing over us, a role that includes controlling activities that pose a threat. That the government's police power includes the authority to control rights incident to property was settled in this country by the Supreme Court in its landmark 1887 decision *Mugler v. Kansas*.[67] In *Mugler*, the Court held that a state could pass a prohibition law that rendered a brewer's investment in his plant and equipment worthless. Yet this law did not, the Court ruled, constitute a taking. For,

> all property in this country is held on the implied obligation that the owners' use of it shall not be injurious to the community. . . . A prohibition simply upon the use of property for purposes that are declared, by valid legislation, to be injurious to the health, morals or safety of the community, cannot, in any just sense, be deemed a taking or an appropriation of property for the public benefit.[68]

Note that Kansas did not take the brewing equipment itself—that remained Mugler's property. Rather, *Mugler* holds that the right to *use* certain property to brew beer does not vest in the owner of that equipment. It vests instead in the public, which can choose to deny it to the owners of the equipment (or anyone else), or as is now the case choose to regulate it in less restrictive ways. In the terms of our argument, *Mugler* affirms the criterion that where rights over property bear closely on the public good, distributive justice places control over them with the public, not with the property's owners.

V

Following Walzer we have determined to seek the just distribution between individual owners and the public of certain rights in the meaning of the good at issue, land. But we have explored three meanings of land, each firmly grounded in our traditions and our legal practice, each with a clear moral appeal. Where then does justice lie? Our commitment to the rule of law seems to force us into a choice: which meaning has the best claim to supporting the most just distribution? But, I suggested at the outset, the three are incommensurable. We are attracted to each for different reasons, and it is unlikely that any one of these reasons would admit to being less compelling than any other. Land as fungible property is at the core of our economic life; it allows for efficient exchanges that increase our society's wealth. Land as personal property speaks to our deep desire for liberty, in the negative sense of protection from governmental intrusion, and in the positive sense of the opportunity for the development of individual identity. And land as community expands our recognition of how our actions can harm other people, hence our sense of connection with them, through nature.

It is wrong, then, to imagine that, for the sake of having consistent takings decisions, we could long sustain a commitment to a particular meaning as definitive. Say we choose land as community, allowing the public to regulate those land use rights it deems likely to cause harm to others. Now consider this circumstance: a couple discovers that the meadow they purchased for a retirement home has been classified as an undevelopable wetland, and their investment in the property is effectively wiped out. Say we choose land as personal property, making it very hard for the public to acquire land use rights. Now consider this circumstance: an entrepreneur seeks to open an outdoor rock music club next to a wildlife refuge, and challenges the local zoning regulations as an infringement of his or her rights of personal expression. Finally, say we choose land as fungible property, so that the public can acquire land use rights on payment of compensation. Now consider this circumstance: a town seeks to regulate construction along the river it uses for its water supply, but to compensate all affected property owners it must raise taxes to an extent that will damage its economy.

Each of these cases is, of course, oversimplified. But each illustrates our inability to make policy on the basis of one meaning alone: situations will inevitably arise where the meaning we have chosen, and its associated distribution, mandates a decision that strikes us as wrong, in particular as less just than would have been mandated by one of the meanings we rejected. To the extent that we abide by our choice, we are sure to be faced with regret. For, the power of each of the meanings is such that when we choose not to honor it, we sacrifice something we value.

It is in this sense that I regard the takings problem as tragic. For Martha Nussbaum, tragedy involves the failure of a protagonist to acknowledge that his or her own value commitment does not exhaust all values.[69] Thus, in her reading of Aeschylus's *Agamemnon*, the tragedy is not simply that Agamemnon sacrificed his daughter—after all he had to, to do the will of Zeus. Rather, he did so without remorse, without acknowledging the weightiness of his violation of familial duties.[70] The play suggests that he erred—and began a cycle of family violence that claims him, his wife, and (temporarily) the sanity of his son—because he did not recognize the multiplicity of moral claims on him, and that he should at least fully face the conflict of his inconsistent obligations.

Conflict is, of course, explicit in Sophocles's *Antigone*. Here, Nussbaum follows Hegel's famous reading, arguing that what marks the two antagonists is their refusal to acknowledge that the other also acts on behalf of a legitimate value.[71] Creon is right to uphold the value of loyalty to the city, wrong to dismiss the claims of piety; Antigone is right to uphold piety and wrong to dismiss her civic obligation. The tragedy is that each holds firm to a simplified vision. By focusing on a single ideal, they both blind themselves to other valid moral claims.

A tragic situation, then, to adopt Hegel's insight, involves a conflict between equally worthy values.[72] The incommensurability of the meanings of land we have surveyed thus gives the takings issue a tragic dimension. If we decide we must, as a matter of policy, protect public rights to environmental quality, we sacrifice something of individual rights; if our policy is to protect individual property, we sacrifice something of the public good. Now it might seem that we could avoid tragedy by recognizing the tragic danger of committing ourselves to a single meaning. We might follow the advice Nussbaum says Tiresias gives to Creon: to adopt "a practical wisdom that bends responsively to the shape of the natural world, accommodating itself to, giving due recognition to, its complexities."[73] That is, we might grant judges the latitude to decide takings cases in light of the particular circumstances they present, appealing to whatever meaning best fits the details of the situation. Indeed, Radin argues that actual takings jurisprudence is best understood "if we see the courts as engaged in the pragmatic practice of situated judgment in light of . . . the unique particularities of each case."[74] As Radin notes, in one landmark takings case, the Supreme Court declared that "takings decisions are 'essentially ad hoc.' "[75]

To adopt an ad hoc approach to takings cases might well mitigate the tragic character of the issue, but will not eliminate it entirely. For even in a single case, especially in a judicial context, there will be a winner and a loser. In every particular decision the price of honoring the meaning that seems most fitting is the need to discount the others. Sometimes it will be easy to decide; other times it will not. There will be certain cases where we will face regret whichever meaning we choose to guide us. In general, that is, each decision involves a concession, based on the need to make a trade-off in this particular situation between values that in general are all attractive.

Further, the ad hoc approach seems to violate our commitment to the political ideal of the rule of law.[76] This ideal holds that our conduct should be governed by general rules, applicable to all, rather than by the arbitrary decisions of powerful individuals. The ideal has clear intuitive appeal: it secures us from the specter of arbitrary authority, wielded unfairly on behalf of some at the expense of others. That is, in John Rawls's words, "the rule of law is obviously closely related to liberty."[77] The connection is made forcefully by Rousseau, for whom the rule of law (in the form of the sovereignty of the general will) is the bulwark against the pinnacle of injustice, the dehumanizing dependence of one person on another characteristic of slavery.[78] Allowing judges to make ad hoc decisions thus seems to threaten our liberty. Would it not be better, we might wonder, to ignore the risk of tragedy implicit in the demand that our legal system embrace a single meaning, for the sake of preserving the liberty encompassed in the rule of law?

In Nussbaum's reading, the lesson of tragedy is that to ignore it is to invite it. *Antigone* shows what happens to "attempts to close off the prospect of conflict and tension by simplifying the structure of the agent's commitments": the agent is destroyed.[79] Thus it is a mistake, in my view, to try to reconceive the takings issue in a way that allows one meaning to be seen as preeminent.[80] Yet I fully accept the importance of the rule of law. Is it possible, then, to reconcile a commitment to the rule of law with the tragic understanding of takings? I believe so. To conclude, I will suggest how that reconciliation might be pursued.

Tragedy is threatened when meanings are incommensurable. But, as Brian Barry shows, incommensurability of values does not lead necessarily to arbitrariness in decision making. Recall that in different situations, we might decide takings cases on the basis of different trade-offs between the meanings we value. Where the harm to the public is greater, we are more willing to sacrifice individuals' control over land; where personal interests are particularly strong, we might accept greater restraints on the government, even at the cost of some environmental damage. But, as Barry observes, such trade-offs can fall into fairly consistent patterns, so that where circumstances are similar, decisions will be based on similar trade-offs.[81] He uses the economists' device of an "indifference map" to express this notion—a set of curves that symbolize all the particular "combinations" of meanings (i.e., "how much" of each would be traded-off for "how much" of the others) that we would judge to be equally acceptable.[82] Say, then, that takings decisions fit this kind of consistent pattern— that is, that where the situations were similar, judges balanced the meanings at stake in similar ways. If this were so, our concern about the rule of law would be greatly obviated, since the legal system would be obeying "the precept that similar cases be treated similarly."[83]

We can use the notion of an indifference map to reinterpret the project of searching for distributive justice in the meanings of goods. The rationale for that project is to link our understanding of justice quite closely to the way the members of society express their understanding of themselves through the way they live. Some goods, like land, have multiple meanings, hence no single authoritative distributive criterion. For the interactions of social life to be ongoing, the members of society are forced to balance their commitments to each meaning in light of particular circumstances. The whole set of these trade-offs exhibits the contours of the dynamic relationship between the meanings of the good that characterizes the given society. The indifference map thus expresses the range of distributions this society would consider just: given this pattern of meaning, these alternative combinations of criteria are equally acceptable.

On this interpretation, an indifference map symbolizes characteristic social practice: the stable pattern that emerges as people trade off their commitments to particular meanings in the interest of sustaining social cooperation. That stability allows us to see that the ideal of the rule of law can be maintained, even where no single meaning is preeminent.[84] And this interpretation allows full rein to the tragic understanding of the takings problem. For Hegel, the source of tragedy is that while our full ethical obligation is to the whole complex of our values, situations arise where, in our actions, we must commit ourselves to only one particular value, a part of the whole. Thus we are led into conflict with those who, committing themselves in action, commit only to other particular values, other parts of the whole. The tragic understanding requires us to see both the ideal whole, and the necessity that in actual situations, the whole must be divided into conflicting particulars.[85] To think of the meanings of land as I now suggest allows us to see them as a totality, as constituting a space of value. But that vision displays them also as particulars, in the particular relationship defined by the characteristic balancing of values society strikes within that space.[86]

Notes

1. The relevant provision of the law is Section 404 of the Federal Water Pollution Control Act of 1972 (33 U.S.C. § 1344).

2. See Ernst Freund, *The Police Power: Public Policy and Constitutional Rights* (Chicago: Callaghan & Co., 1904), § 511.

3. See Timothy Egan, "Unlikely Alliances Attack Property Rights Measures," *New York Times*, 15 May 1995, A1.

4. For recent overviews of the ecology and regulation of wetlands see National Research Council, *Wetlands: Characteristics and Boundaries* (Washington: National Academy Press, 1995) and Mark S. Dennison and James F. Berry, eds., *Wetlands: Guide to Science, Law, and Technology* (Park Ridge, N.J.: Noyes Publications, 1993).

5. I have cast this argument in straightforwardly anthropocentric terms: justice is presented as a matter of human interests. In this paper I shall pursue an anthropocentric course. But it is certainly true that arguments for the justice of environmental regulations can be cast in nonanthropocentric terms, with reference to the rights of other animals or other living creatures, or the rights of the environment as a whole.

6. This question was first articulated by Aristotle, whose principle of distributive justice is that equals should be treated equally (*Nichomachean Ethics*, v.3). In our day, the central work in the field is John Rawls's *Theory of Justice* (Cambridge: Harvard University Press, 1971) (henceforth TJ), which defends what he calls the "difference principle": inequalities are justified if the worst off in society are better off with them than if they did not exist.

7. See Bruce Ackerman, *Private Property and the Constitution* (New Haven: Yale University Press, 1977), 26 ff.; A. M. Honoré, "Ownership," in *Oxford Essays in Jurisprudence*, ed. A. G. Guest (Oxford: Oxford University Press, 1961).

8. Libertarian opponents of zoning believe that this distribution of rights is unjust. They hold that all the rights incident to ownership, including development rights, vest in the title holder, unless surrendered in voluntary (usually market) exchanges (Ellen Frankel Paul, *Property Rights and Eminent Domain* [New Brunswick: Transaction Books, 1987], 146–7 [henceforth PR], citing the work of Bernard Siegan and Robert Ellickson).

9. *Spheres of Justice* (New York: Basic Books, 1983), 86 ff.

10. "To Each His Own" (review of *Spheres of Justice*), *New York Review of Books*, 14 April, 1983, 4–6.

11. " 'Spheres of Justice': An Exchange," *New York Review of Books*, 21 July, 1983, 44.

12. *Justice and Interpretation* (Cambridge: MIT Press, 1993), 20. Walzer himself says very little about this issue. He briefly notes that bread, for instance, means among other things the staff of life and the body of Christ, but only to support his claim that no one meaning is primary in all societies (*Spheres of Justice*, 8). But, to my knowledge, he never grapples with the case of a single good supporting divergent meanings within a given society at a given time.

13. See Margaret Radin, "Reconsidering the Rule of Law," *Boston University Law Review* 69 (1989): 781–819.

14. This point was made by Bentham in his utilitarian account of property rights (*Principles of the Civil Code* [London, 1864], pt. 1, chaps. 8 and 10), and is endorsed by John Rawls in his discussion of the rule of law (TJ, 235–36).

15. The administration of a property regime involves not only the decisions of judges—clearly the content of legislation and the actions of administrative agencies also have crucial effects on the content of property rights.

16. See Rawls, who discusses the rule of law as an "elucidation" of the priority of the principle of liberty among the other principles of justice (TJ, 235).

17. This follows from a broad suggestion by Amy Gutmann to "look for moral considerations that can adjudicate among conflicting meanings." Her argument goes on, however, to urge leaving meanings behind. She construes the multiplicity of meaning as a sign that people in society have moral disagreements about the good in question. Resolving such disagreements requires moral argument, but in conducting moral arguments "both [sides] must move beyond social meaning," to reason instead about "which [distributive] policy entails the lesser moral wrong" ("Justice Across the Spheres," in *Pluralism, Justice, and Equality*, ed. David Miller and Michael Walzer [Oxford: University Press, 1995], 99, 110).

18. See *Political Argument: A Reissue with a New Introduction* (Berkeley: University of California Press, 1990), 4–8 (henceforth PA).

19. *Market, State, and Community: Theoretical Foundations of Market Socialism* (Oxford: Oxford University Press, 1989), cited by Barry as an improved statement of his own view (PA, xl).

20. I am adopting the Hegelian conception of tragedy as the conflict between two goods. For a clear exposition, see A. C. Bradley, "Hegel's Theory of Tragedy," reprinted in *Hegel on Tragedy*, ed. Anne Paolucci and Henry Paolucci (New York: Harper Torchbooks, 1975).

21. The term "intuitionistic" is Rawls's; see TJ, 34 ff.

22. Margaret Jane Radin, "The Liberal Conception of Property: Cross Currents in the Jurisprudence of Takings," *Columbia Law Review* 88 (1988): 1685.

23. See "Property and Personhood," *Stanford Law Review* 34 (1982): 986–88, 991–1008.

24. In the words of James W. Ely, Jr., "It is difficult to overstate the impact of the Lockean concept of property" (*The Guardian of Every Other Right: A Constitutional History of Property Rights* [Oxford: Oxford University Press, 1992], 17). See also Eugene C. Hargrove, "Anglo-American Land Use Attitudes," *Environmental Ethics* 2 (1980): 121–48; Walton H. Hamilton, "Property—According to Locke," *Yale Law Journal* 41 (1932): 864-80.

25. *Second Treatise of Government* (Indianapolis: Hackett, 1980), § 34 (further citations will be made in the text).

26. One must also, of course, leave "enough, and as good" for others (§ 33).

27. See *The Political Theory of Possessive Individualism: Hobbes to Locke* (Oxford: Oxford University Press, 1962), 221–38.

28. "Liberal Conception," 1686.

29. Perhaps the most important contemporary retelling of the Lockean

story is found in Robert Nozick, *Anarchy, State, and Utopia* (New York: Basic Books, 1974), 150 ff.

30. *Takings: Private Property and the Power of Eminent Domain* (Cambridge: Harvard University Press, 1985), 61.

31. Radin, "Property and Personhood," 992.

32. "Reconstructed but Unregenerate," in *I'll Take My Stand: The South and the Agrarian Tradition*, by Twelve Southerners (Baton Rouge: Louisiana State University Press, 1977), 18.

33. *I'll Take My Stand*, 19–20. See, also, the rather idealized description of rural life in Andrew Nelson Lytle's contribution, "The Hind Tit."

34. See Alexander Karanikas, *Tillers of a Myth: Southern Agrarians as Social and Literary Critics* (Madison: University of Wisconsin Press, 1966), chap. 3. Karanikas notes the Jeffersonian inspiration of agrarian thought on property (47).

35. *The Southern Agrarians* (Knoxville: University of Tennessee Press, 1988), 171.

36. Ransom, *I'll Take My Stand*, 19.

37. These implications of the agrarian attitude are explored more fully in the writings of Wendell Berry.

38. This is the reasoning behind the view that the "tragedy of the commons" is best avoided by apportioning common resources into individually owned holdings—a view that dates back to Aristotle (*Politics*, book II, chap. 5, 1263a21).

39. For general discussions of the connection between liberty and property see Alan Ryan, *Property* (Minneapolis: University of Minnesota Press, 1987), chaps. 3, 7; and Andrew Reeve, *Property* (London: Macmillan, 1986), chap. 4.

40. "American Constitutionalism and the Paradox of Private Property," in *Constitutionalism and Democracy*, ed. Jon Elster and Rune Slagstad (Cambridge: Cambridge University Press, 1988); see also *Private Property and the Limits of American Constitutionalism: The Madisonian Framework and Its Legacy* (Chicago: University of Chicago Press, 1990).

41. Nedelsky, "American Constitutionalism," 263.

42. Statement by Fairness to Land Owners Committee, in U.S., Congress, House, Committee on Merchant Marine and Fisheries, *Takings, Compensation, and Pending Wetlands Legislation: Hearing on the Future Course of the Federal Wetlands Program*, 102nd Cong., 2d sess., 21 May, 1992, 74–5.

43. Hearing, 2. See also the statement of the Pacific Legal Foundation, 89. This notion has a long heritage, and was stated forcefully at the time of the American Revolution by Arthur Lee: "The right of property is the guardian of every other right, and to deprive a people of this, is in fact to deprive them of every other liberty" (quoted in Ely, *The Guardian of Every Other Right*, 26).

44. Hargrove, "Anglo-American Land Use Attitudes," 131.

45. Though as Stanley Katz observes, Jefferson did not regard property in land as an end in itself, but rather justified it as instrumental to full democratic

citizenship ("Thomas Jefferson and the Right to Property in Revolutionary America," *The Journal of Law and Economics* 19 [1976]: 467–88).

46. PR, 90.

47. PR, 32.

48. PR, 90. In this light, FLOC seems to concede too easily when they demand compensation by fair market value only. If its members' property means as much as their aggrieved rhetoric indicates, it would seem that virtually no amount of compensation could make good their loss.

49. PR, 254–60. Note that she argues for this claim by appealing to rights, rather than to the notion of the personal meaning of property.

50. See PR, chap. 1 for her criticism of environmentalism.

51. *A Sand County Almanac and Sketches Here and There* (Oxford: Oxford University Press, 1987), 204 (further citations will be made in the text).

52. Note that Leopold was by no means opposed to private ownership of land—indeed, he addresses the prescriptions of the land ethic precisely to private land owners (213–14).

53. Leopold suggests that this recognition can lead to a life more fulfilling for being lived in community with the land; this is the lesson of the year of life he records in *A Sand County Almanac*.

54. For an account of the development of the community notion within the science of ecology see Gregg Mitman, *The State of Nature: Ecology, Community, and American Social Thought, 1900–1950* (Chicago: University of Chicago Press, 1992).

55. For example, Rachel Carson's *Silent Spring* (Boston: Houghton Mifflin, 1962) played a decisive role in popularizing the idea that natural interrelations propagate utterly unforeseen damages from one element of an ecosystem (crop lands sprayed with DDT), to what are revealed as only apparently disconnected elements (song birds).

56. National Research Council, *Wetlands*, 215. Economists have explored techniques for determining the financial value of the services wetlands provide; see, for example, Stephen Farber and Robert Costanza, "The Economic Value of Wetlands Systems," *Journal of Environmental Management* 24 (1987): 41–51.

57. Leopold's ethical prescription was given legal force in the well-known Wisconsin takings decision *Just v. Marinette County*, 201 N.W. 2d 761 (1972). In that case, the Wisconsin Supreme Court held that the owners of a lakeside wetland had no right to alter the natural character of their land (by filling it), because "the changing of wetlands and swamps to the damage of the general public by upsetting the natural environment and the natural relationship is not a reasonable use of land," hence could be legally prevented by regulation (768).

58. On this point I disagree with J. Baird Callicott (in "Hume's Is/Ought Dichotomy and the Relation of Ecology to Leopold's Land Ethic," *Environmental Ethics*, 4 [1982]: 163–174) and with Holmes Rolston III (in "Is There an

Ecological Ethic," in *Philosophy Gone Wild: Environmental Ethics* [Buffalo: Prometheus Books, 1989]), who both argue that from the "is" of ecology Leopold derives a squarely nonanthropocentric "ought."

59. The study of the human use—and attendant transformation—of the land is central in the classic geographical work of L. Dudley Stamp—see, for example, *Applied Geography* (Harmondsworth: Penguin, 1960).

60. "Takings, Private Property and Public Rights," *The Yale Law Journal* 81 (1971): 155.

61. "Takings," 155.

62. Because of easily recognizable collective action problems: free riding among affected parties, insufficient motivation among individuals, and high transaction costs (Richard A. Posner, *Economic Analysis of Law*, 4th ed. [Boston: Little, Brown and Co., 1992], 63).

63. "Takings," 160.

64. "Takings," 159.

65. Sax puts this point in terms of maximizing benefits; see "Takings," 158, 160, 186.

66. This pattern is clearest in Rousseau, *The Social Contract*, book 1, chaps. 6, 9; it can also be seen in Locke's view that in joining society, individuals surrender to the public their power of enforcing the law of nature (§ 89).

67. 123 U.S. 623.

68. 123 U.S. 665, 668–69, emphasis added.

69. She presents her view in *The Fragility of Goodness: Luck and Ethics in Greek Tragedy and Philosophy* (Cambridge: Cambridge University Press, 1986), esp. chaps. 2–3 (henceforth FG).

70. FG, 32–38.

71. FG, chap. 3.

72. *Hegel on Tragedy*, 132.

73. FG, 80. Cf. Rawls on intuitionism, TJ, 34 ff., discussed by Barry, PA, xxxix ff.

74. "Diagnosing the Takings Problem," in *Compensatory Justice* (NOMOS XXXIII), ed. John W. Chapman (New York: New York University Press, 1991), 270. Radin attributes the lack of general principles in takings cases to, among other things, the tension between the fungible and personal conceptions of property, and the impossibility of establishing in principle whether a particular entitlement is of one sort or the other (259). Carol M. Rose offers an analysis of the takings "muddle" with a similar logical structure; for her, courts are torn between a conception of property as a means to aquire wealth, and as a means to participate in civic life ("Mahon Reconstructed: Why the Takings Issue Is Still a Muddle," *Southern California Law Review* 57 [1984], 561–99).

75. "Diagnosing," 277, n. 60.

76. See Radin, "Reconsidering the Rule of Law."

77. TJ, 235. Note that, for Rawls, the ideal of the rule of law is connected to his rejection of the notion that values are incommensurable, in favor of a strict priority of one value—liberty—over all others in his conception of justice.

78. See Zev Trachtenberg, *Making Citizens: Rousseau's Political Theory of Culture* (London: Routledge, 1993), 213–17, citing, for example, *The Social Contract*, bk. 1, chap. 7.

79. Nussbaum, FG, 51.

80. This effort might take the form of a libertarian assertion of the preeminence of individual property rights, or an extreme environmentalist demand that nature be protected at any cost.

81. See PA, 4–8.

82. The quantitative language is intended to be merely suggestive. While it is foolish to think of specific "amounts" of each meaning, it does make sense to think of a meaning being implemented to a greater or a lesser degree (see Barry, PA, 6).

83. Rawls, TJ, 237. I am not here defending the claim that our takings jurisprudence does in fact fall into the kind of pattern I describe.

84. Indeed, Radin suggests that it is best to reinterpret the ideal of the rule of law, by adopting a Wittgensteinian conception of rules as emerging out of a form of life ("Reconsidering the Rule of Law," 797–800).

85. *Hegel on Tragedy*, 132.

86. Portions of the research for this paper were supported by a University of Oklahoma Junior Faculty Summer Research Grant. My thanks to Mike Koessel, Gregg Mitman, Don Pisani, Andrew Light, and the anonymous reviewers for this journal for their helpful comments and suggestions; my misunderstandings are my own.

Muslim Contributions to Geography and Environmental Ethics: The Challenges of Comparison and Pluralism

James L. Wescoat, Jr.

This paper arises from a concern that environmental ethics, as pursued in the United States and Europe, does not comprehend the contributions of other culture groups and regions of the world, past or present. A deeper concern is that philosophical research in geography and environmental ethics does not speak to the condition of most of the world's peoples or places. As a step toward broadening the conversation, this chapter surveys Muslim contributions to geography and environmental ethics.[1]

The overlapping terrains between geography and environmental ethics are explored here in three ways: (1) recognition of Muslim contributions to geographic and environmental ethics research; (2) identification of intersections between geography and environmental ethics in Islamic thought; and (3) brief discussion of initiatives that seek to communicate across cultural-geographic realms and contribute toward a constructive pluralism (fig 5.1).[2]

I begin with a simple definition of "pluralism" as a situation characterized by apparent differences in the form and content of practical reasoning.[3] Constructive pluralism implies a commitment to coexistence, at the least, to respectful efforts at communication in the near term, and ultimately toward creative cooperation. Although comparative geographical research has not always served these aims, and

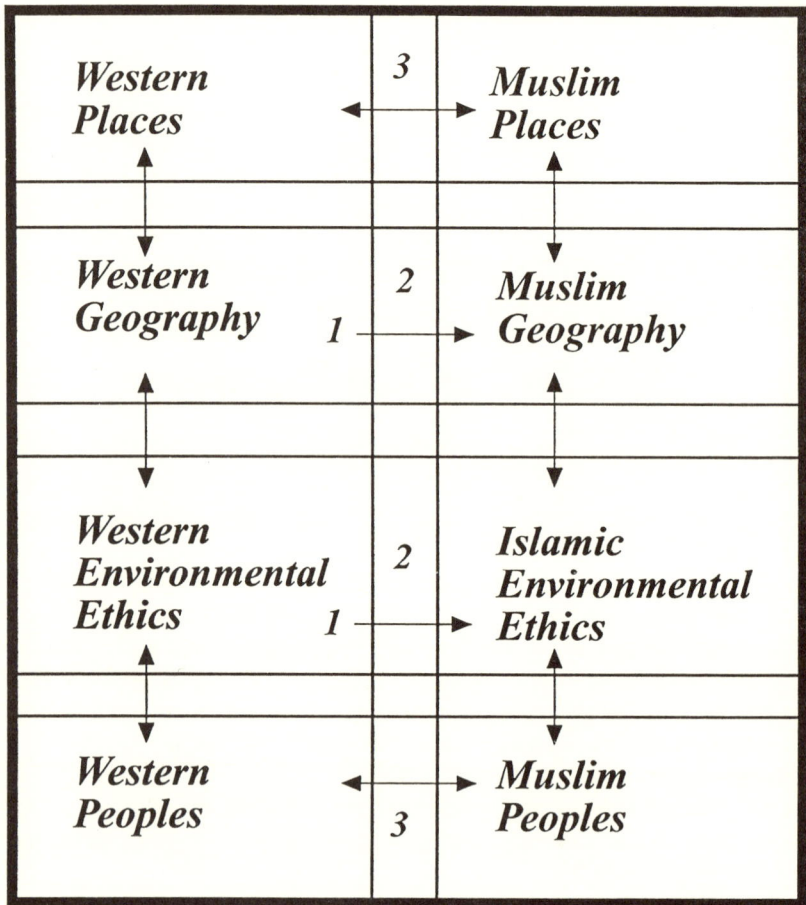

Fig. 5.1. Overlapping Terrain I

remains theoretically and methodologically underdeveloped, I hope to demonstrate how it might take a more constructive path.[4]

I focus on Muslim contributions for three reasons. First, they have a significant record of contribution to both geography and ethics. Second, they pose some of the most serious challenges for comparison and pluralism in the United States.[5] Third, my research in a Muslim country affords me some firsthand experience in trying to understand and address those challenges.[6]

The method of my study involves mapping the relations indicated in figure 5.1. The first section identifies patterns and gaps in Western awareness of Muslim geography and environmental ethics. The second section fills in some of those gaps by exploring overlapping terrain among geography, ethics, and environmental concerns in early Islamic thought. Demonstrating that gaps exist between the inquiries of one society and the awareness of another, and that those gaps can be filled, is not inherently useful. It is also necessary to show how those gaps might undermine a constructive pluralism, which, if fostered, could speak to contemporary environmental and social problems. Therefore, the final section of the chapter describes recent international efforts to bridge, communicate, or otherwise understand the contributions of Muslim societies, places, and scholars.

Recognition

In their comparative study of minority group aspirations, Mikesell and Murphy identify recognition as the most basic aim of social groups.[7] The same may be said of small academic fields, such as environmental ethics and the philosophy of geography, which seek recognition from their respective disciplines as well as from each other (fig 5.2). Focusing on the disciplinary terrain of these fields has a narrowing effect, however, when it draws attention from other large cultural groups, traditions, and regions of the world.

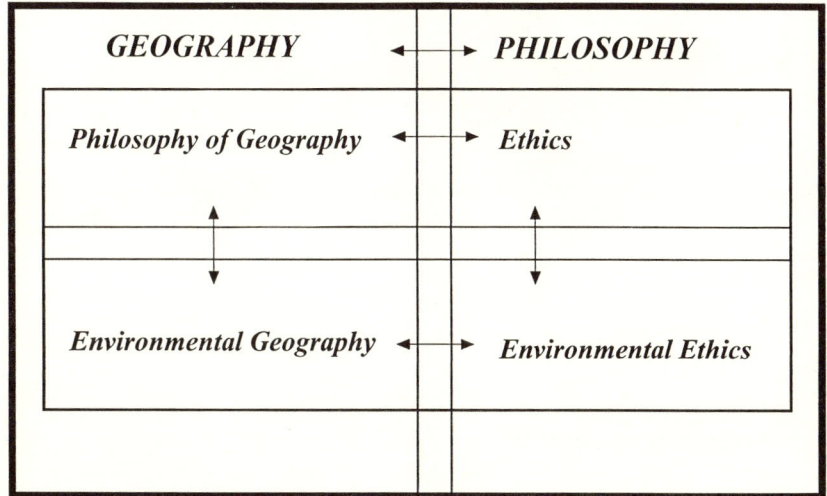

Fig. 5.2. Overlapping Terrain II

Muslim geography and ethics are also small academic fields, but they are situated within a major array of religious and societal groups that involve large numbers of people, a long historical record of accomplishments, and vast geographic areas. Figure 5.3 indicates countries that have a Muslim majority or large minority populations. The present population of Muslims has been estimated at more than one billion, or about 18 percent of the world population.[8] Most countries have sizable or growing Muslim minorities. The number of Muslim groups in the United States, for example, is growing through conversion, immigration, and natural increase.[9] The number of Muslims living in the United States was estimated at 3.3 million in 1980 and projected to be 4 million in 1986.[10] This included African-Americans (30.2 percent) as well as Middle East/North Africans (28.4 percent), Eastern Europeans (26.6 percent), and Asians (11.5 percent).[11]

Despite these impressive figures and images, surprisingly little attention has been given to the geographic contributions of Muslim peoples in contemporary Western scholarship.[12] A rough indication may be

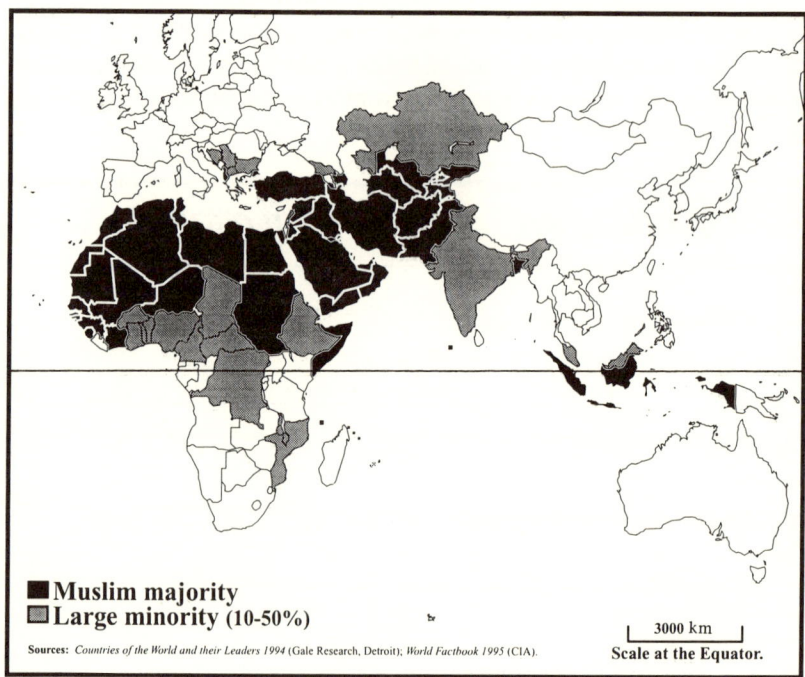

Fig. 5.3. Countries with Muslim Population

gained from bibliographic citations to geography and ethics research in Muslim contexts (table 5.1). In the *GEOBASE* index of geographical research for 1990 to 1995, for example, the intersection of geog* + environ* + ethic* is small (13). But works on Islam* + geog + environ*, or Islam* + environ* + ethic* are virtually nil. Geographic research on Islam rarely involves environmental issues. Turning to *The Philosopher's Index* for a much longer period (1940–1995), one finds that "Islam" figures more prominently than "geography." But neither of them is associated with environmental ethics (0 hits), or with each other (1 hit). The new four-volume *Oxford Encyclopedia of the Modern Islamic World* has articles on "ethics" and "gardens" but not "geography," "environment," or "ecology."[13]

These omissions in late-20th century compilations reflect the vicissitudes of geographical research in the United States as well as in Muslim regions as compared, at least in the latter case, with brilliant periods and centers of both geographic and philosophical research in the tenth through sixteenth centuries. Thus, as a starting point, this chapter surveys early contributions of Muslim scholars to geography and ethics, which have direct or indirect concern with "environmental ethics."[14] I hope to show that the overlapping scholarly terrain between geography and environmental ethics, past and present, has continuing relevance for both Muslim and non-Muslim peoples, as well as the relations between them.[15]

On the face of it, the recent failure to recognize Muslim contributions constitute a step backward from earlier academic coverage in works like the *Encyclopedia of Islam*, which includes detailed entries on geography (*djughrafiya*), this world (*dunya, hayat*), and environmental topics (e.g., water [*ma*] and animals [*dabba, hayawan*]). However, critics remind us that "coverage" is not inherently good, from a moral or political standpoint, for it can undermine or subvert, rather than enable, a constructive pluralism.[16]

As with any topic, databases and encyclopedias may miss important scholarly contributions by less-cited authors and publications. Most serious, the citations in table 5.1 do not include works by scholars or publications from the Middle East or South Asia.[17] But even these works are few in number, which reflects the limited scholarly activity in these fields today.

To compound these patterns of scholarly neglect, Islam has special recognition problems in the United States. Perceptions of Islam and Muslims are uninformed in many respects, misinformed in others, and pejorative in ways that have aggravated confusion and conflict. In light

TABLE 5.1
Citations To Research On Islam, Geography, and Environmental Ethics

The Philosophers Index (1940–95)		GEOBASE Index (1990–95)	
Ethic*	−224,008	Ethic*	−281
Environ*	−1,783	Environ*	−41,381
Geog*	−151	Geog*	−19,335
Environ* + Ethic*	−281	Environ* + Ethic*	−281
Geog* + Ethic*	−56	Geog* + Ethic*	−56
Geog* + Environ*	−14	Geog* + Environ*	−14
Geog* + Environ* + Ethic*	**=0**	**Geog* + Environ* + Ethic***	**=13**
Islam*	=551	Islam*	=392
Islam* + Ethic*	=89	Islam* + Ethic*	=2
Islam* + Ethic* + Environ*	=4	Islam* + Ethic* + Environ*	=0
Islam* + Geog*	=1	Islam* + Geog*	=33
Islam* + Geog* + Environ*	=0	Islam* + Geog* + Environ*	=1

Source: CD-ROM citation indexes in the University of Colorado Libraries.

of these problems, it seems helpful to focus on the under-recognized contributions of Muslims to geography and environmental ethics as a path toward constructive comparison and pluralism in those fields.

Overlapping Terrain

Muslim contributions to environmental ethics may be discerned, in part, by mapping the varieties of scholarship in geography and ethics and then identifying the intersections among them (fig 5.4).[18]

Varieties of Muslim Geography

Muslim geographic inquiry has ranged across many genres, from the astronomical and speculative to the descriptive and literary.[19] Table 5.2 groups these works under nine broad headings. Although all of them have had some connection with geographic inquiry, only a few

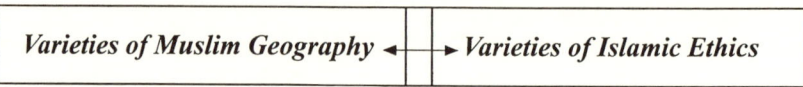

Fig. 5.4. Overlapping Terrain III

TABLE 5.2
Varieties Of Muslim Geographic Science

1. Astronomical and cosmographical studies
2. World geographic descriptions—of climates (*iqlim*), regions (*kishwar*), and countries (*baldan, mamalik*) of the world (*ard*)*
3. Geographic dictionaries—lists of regions, towns, rivers, and other topographic categories, sometimes with coordinate locations
4. Geographic sections of broader treatises on wisdom (*hikma*), science (*ilm*), and philosophy (*falsafah*)—e.g., the *Epistles* (*Rasa'il*) of the Sincere Brethern of Purity (*Ikhwan al-Safa*); and al-Biruni's *India*
5. Geodesy and maritime literature—for commercial and political purposes
6. Regional description
    a. Administrative areas for revenue collection and governance
    b. Topographical literature—compiling information about important sites and facts about an area for general purposes, including plants and animals*
    c. Agronomic literature—practical sciences of cultivation and agricultural diffusion
    d. Cultural surveys—e.g., al-Biruni and the *Ain-i-Akbari*
7. Travel literatures
    a. Routes and stages for military travel
    b. Trade and travel guides for merchants
    c. Pilgrimage (*ziyarat*) stories and guides*
    d. Travel accounts of individual adventurers*
8. Modern geographic inquiry of the 19th and 20th centuries (heavily influenced by Western colonial and post-colonial geography)
9. Geographic references in other genres
    a. Autobiography/memoirs
    b. Biography
    c. Histories
    d. News and intelligence reports
    e. Poetic and mystical literature

Sources: R. Ahmed, "Djughrafiya," in *Encyclopedia of Islam*, new edition (Leiden: E. J. Brill, 1979); al-Biruni. *Al-Beruni's India*, trans. E. Sachau (Lahore: Sheikh Mubarak Ali, 1963 [reprint]); K. W. Butzer, "The Islamic Traditions of Agroecology: Cross-Cultural Experience, Ideas and Innovation," *Ecumene: A Journal of Environment, Culture, Meaning* 1 (1994): 7–50; D. F. Eickelman and J. Piscatori, *Muslim Travellers: Pilgrimage, Migration, and the Religious Imagination* (Berkeley: University of California Press, 1990); A. T. Karamustafa, "Cosmographical Diagrams," in *Cartography in the Traditional Islamic and South Asian Societies*, vol. 2 in *The History of Cartography*, eds., J. B. Harley and D. Woodward. (Chicago: University of Chicago Press, 1992), 71–89; A. T. Karamustafa, 1992b. "Military, Administrative, and Scholarly Maps and Plans." In *Cartography in the Traditional Islamic and South Asian Societies*, pp. 209–27; J. H. Kramers, "Djughrafiya," in *First Edition of the Encyclopedia of Islam, 1913–1936. Supplement*, vol. 9 (Leiden: E.J. Brill, 1979), 61–73; G. R. Tibbetts, "Early Geographical Mapping," in *Cartography in the Traditional Islamic and South Asian Societies*, pp. 90–105; Yaqut al-Rumi, *The Introductory Chapters of Yaqut's Mu'jam al-Buldan*, tr. Wadie Jawaideh (Leiden: E.J. Brill, n.d.).
*Categories relevant to environmental ethics.

have been explicitly regarded as *djughrafiya*. The latter have included massive geographic dictionaries, works of geodesy, cosmology, scientific cartography, and certain types of regional description. They required large-scale patronage of the sort marshalled by the Arab caliphate in Baghdad from the ninth through twelfth centuries and somewhat later in the Central Asian khanates of Balkh, Bukhara, Samarqand, and Khorezm. [20] Smaller centers of patronage arose in Spain, Sicily, Egypt, and Afghanistan.

A third wave of "sponsored geographic research" occurred in the so-called gunpowder empires of the sixteenth and seventeenth centuries in Ottoman Turkey and, to a lesser extent, Safavid Persia and Mughal India.[21] Some brilliant work, often of a topographic nature, occurred at regional courts and centers across the Islamic realm.[22] Because imperial geographic patronage was more often political and economic than religious in aim, its products may be regarded as "Muslim" (cf. "Islamic" pilgrimage and prayer maps and charts).[23]

Although much Muslim scientific work is prefaced by religious invocations and praise, only a few genres had primarily religious aims. Pilgrimage literature linked worship, routes, and places—as did some poetry and mystical writing about the world. Cosmologists pursued abstract theological issues. Some topographic works concentrated on the distribution of religious endowment lands (*auqaf*) or saints' shrines (*mazars*). Each genre, including the more secular ones, influenced environmental understanding and norms.

The boundaries among these categories should not be drawn too sharply. In principle, Islam is a complete system of life that subsumes worship, thought, politics, and nature. Sharp distinctions violate this unity and fail to grasp the aims of syncretic works that have sought to combine cosmology, theology, topography, and ethics.[24]

Varieties of Islamic Ethics

Distinctions among the aims, sources, and varieties of ethical inquiry are also tenuous for they draw in varying ways upon scripture, theology, philosophy, mysticism, folklore, and the literatures of norms (*adab*) and virtues (*akhlaq*) (table 5.3). Islamic ethics also subsumed norms derived from pre-Islamic Arabian, Greek, and Persian sources.[25] But they are held together as "Islamic" traditions by a common reliance upon the primary sources of the *Qur'an* and *sunnah*.

Not surprisingly, the connections between ethics and religion are far stronger than those between geography and ethics or geography and

TABLE 5.3
Varieties Of Islamic Ethics

1. Scriptural morality (*Qur'an* and *hadiths*)
2. Religious ethics (traditionalist, legal)
3. Theological ethics (*kalam*—rationalists, voluntarists)
4. Philosophical ethics (*falsafah*—Aristotelian, neo-Platonist, neo-Pythagorean)
5. Sufi orders (devotional, experiential, mystical)
6. Personal and professional ethics (*akhlaq* and *adab*)
7. Modernist Ethics (seeks to account for and move beyond the 19th–20th century cultural encounters with the West)
8. Popular wisdom literature, folklore and poetry (e.g., *Kalila va Dimna*)

Sources: Mehmet Ayedin, "Islamic Ethics," in *Encyclopedia of Ethics,* vol. 1, Becker and Becker, eds., (New York: Garland Publishing, 1992), 631–634; M. Fakhry, *Ethical Theories in Islam* (Leiden: E.J. Brill, 1991); G. F. Hourani, *Reason and Tradition in Islamic Ethics* (Cambridge: Cambridge University Press, 1985); R. G. Hovannisian, ed., *Ethics in Islam* (Malibu: Undena Publications, 1985); G. Kramer, "Minorities: Minorities in Muslim Societies," in *The Oxford Encyclopedia of the Modern Islamic World,* vol. 3 (Oxford: Oxford University Press, 1995), 108–10; B. D. Metcalf, *Moral Conduct and Authority: The Place of Adab in South Asian Islam* (Berkeley: University of California Press, 1984).

religion. But in common with geography, ethics are strongly linked with politics and economics. And efforts were made to integrate both geography and ethics with general philosophic ideas about wisdom (*hikma*), unity (*wahdat*), and law (*shari'a*).

For practicing Muslims, the *Qur'an* and *Sunnah* (the life and sayings of the prophet Muhammad) constitute the primary sources of moral guidance and of *shari'a* law. Theologians distill formal ethical principles from these sources for other theologians, religious leaders (*'ulama*), judges, and lawyers. The latter groups along with local preachers (e.g., *mullahs*) elaborate and enforce practical religious ethics for the community (*ummah*).

Philosophical ethics intersect with religious ethics on matters of virtue and conduct (*akhlaq*). They build on Aristotelian and Platonic foundations with varying commitments to religious ethics, which some philosophers deemed as merely necessary for the masses. However, ethics was a small field of classical Islamic philosophy compared with metaphysics and science.[26] Reflection about the nature of the physical world and animals was more commonly found in these latter fields of philosophy than in ethics.

The devotional works of Sufi mystics cut across social and intellectual classes. Sufism conjoins religious experience and intellectual inquiry. When these sources conflict, Sufis have tended to rely upon the former. They often came into conflict with orthodox religious morality and law. Some Sufis were executed as heretics. Others, perhaps the majority, were of a sober, practical temperament, but they too tended to dismiss philosophical ethics as sterile or false. According to al-Ghazali (1058–1111 C. E.): "The philosophers have taken over this teaching [of pious mystics] and mingled it with their own disquisitions, furtively using this embellishment to sell their own rubbishy wares more readily. . . . When a man looks into their books, such as the 'Brethern of Purity' and others . . . he readily accepts the falsehood they mix with that [truth], because of the good opinion resulting from what he noticed and approved."[27] As will be discussed later, the Brethern of Purity made one of the most important contributions to environmental ethics in the history of Islamic thought. Was it truth, falsehood, or a mixture? A great deal of classical Islamic thought has wrestled with the challenges of synthesizing and critically evaluating scriptural, theological, legal, and mystical traditions.[28]

These efforts were disrupted and made far more difficult by the ruptures and overlays of colonialism in the fields of law, politics, economics, and geography. Although some modernists in Turkey, India, and Egypt have sought to reweave ethical and geographical concerns, they face formidable challenges from traditionalists and from the West. For example, Sayyid Ahmed Khan produced groundbreaking, controversial, works on theology, education, and topographic description in India—stimulated by and reacting to the colonial encounter.[29] Allama Muhammad Iqbal rewove European and Islamic social thought to envision new prospects for Muslim culture and territory in South Asia.[30]

Popular strands of wisdom literature and cautionary tales must not be overlooked. They have had broad appeal from pre-Islamic times to the present. In the story of *Layla and Majnun*, for example, wisdom and madness were associated with an intense love of nature and animals.[31] And the *Kalilah va Dimnah*, in which animals have the main roles in moral fables, has special relevance for understanding the overlapping terrain between environmental and ethical inquiry.[32]

Overlapping Terrain

The intersections between Muslim geography and ethics are thus many but fragmented; their relevance for environmental ethics potentially rich but tenuous. On the first point, it is useful to recall one of the principles that holds the picture together, namely that all of the varieties of ethical and geographic inquiry described above must square, in some way, with the revealed truth of the *Qur'an* and *sunnah*.

Philosophers may dismiss Sufis, and vice versa, but neither may separate environmental concerns from their broader religious context. Muslim authors have yet to envision a separate realm of environmental ethics in the manner of some Western philosophers, scientists, deep ecologists, or Greens.

If we overlay tables 5.2 and 5.3 to discern intersections between Muslim geography and ethics that are relevant for environmental ethics, several important historical experiments stand out. The earliest but least interesting intersection occurs in individual scholars who wrote about both ethics and geography. For example, one of the earliest Muslim philosophers, Yacub al-Kindi (d. 870), wrote both a *Description of the Inhabited Part of the World* based on Ptolemy and a collection of excerpts on ethics (*Alfaz Suzrat*).[33] These texts reflected a dual interest in geography and ethics but not a significant synthesis of them.

A more substantial contribution came from Abu Zaid Ahmed ibn Sahl al-Balkhi (d. 934), after whom the Balkhi school of cartography is named.[34] Al-Balkhi was a student of al-Kindi in Baghdad. In his later years he produced the prototype for what would become the "Islam-Atlas," a collection of maps of the Islamic realm accompanied by descriptions of their physical and social features.[35] Although not a work of ethics as such, the Balkhi school nevertheless established a tradition of explicitly Islamic geography, characterized by its core concern for Muslim places and peoples (*Dar al-Islam*).

One of the most remarkable works connecting geography and environmental ethics dealt with animals. The *Rasa'il* (epistles) of the *Ikhwan al-Safa* (the Sincere Brethern of Purity) were produced by a secret society of the tenth and eleventh centuries in southern Iraq.[36] The brethern were inspired by the "Tale of the Doves" in the *Kalilah wa Dimnah,* noted earlier, in which a large number of doves, seduced by a glimmer of grain set out as bait, were captured under a hunter's net.[37] Each struggled for freedom, in vain; finally their leader organized them to fly as a group in one direction, lifting the net to safety. Those who had not been ensnared helped by pulling on the net from above.

The Sincere Brethern sought to form a loyal community during a time of political instability. Al-Faruqi records one description of them as "a society for ethical culture."[38] Anticipating an escalating role for themselves, they produced a collection of more than fifty epistles dealing with the natural, moral, and cosmological sciences.[39] Here we focus on a portion of one epistle that stands as a major Islamic contribution to environmental ethics, situated within a truly expansive philosophical context.

The epistles proceed, in encyclopedic fashion, from mathematics (fourteen in number) to natural science (seventeen), psychology and logic (ten), and ultimately to metaphysics, law, and theology (eleven).[40] Their use of fables and parables suggests that they served as instructional aids for the practical and esoteric advancement of initiates. The epistles on geography and animals occur in a section on physics that commingles natural and spiritual concerns (table 5.4).

The epistle on animals is titled, "The Case of the Animals Versus Man before the King of the Jinn." The animals lived in peace and harmony prior to their enslavement and abuse by humans. After an episode of violent enslavement, the animals appeal to the king of the *jinn*.[41] The king calls together representatives from all the peoples of the world, including Jews and Christians as well as Muslims, and all of the animal kingdoms of the world. One after another, the animals recount in heartrending detail the gross injustices committed upon them by human beings. These arguments are countered, generally unpersuasively, by human assertions of their superiority and privi-

TABLE 5.4
Epistles (*Rasa'il*) On Physics

1. Matter, form, motion, time, place, and their relations
2. The heaven and cosmos and the refinement of character and soul
3. On generation and corruption
4. On influences from above
5. On the generation of minerals
6. On the essence of nature
7. On the kinds of plants
8. On the generation of animals and their kinds (including *The Case Of The Animals Versus Man*)*
9. On the composition of bodies
10. On sense, perception, and the perceived
11. Embryology (including astral influences)
12. On man as a microcosm
13. On the resurrection of souls and bodies
14. On man's capacity for awareness
15. Life and death
16. Pleasures and pains related to life, death and the soul
17. Causes of the diversity of language and linguistic issues

Source: L. Goodman, trans., *The Case of the Animals versus Man Before the King of the Jinn, A Tenth-Century Ecological Fable of the Pure Brethren of Basra* (Boston: Twayne Publishers, 1978), 46.

leges. Although the king ultimately decides that humans do have a measure of superiority over animals in the creation, the weight of the judgment falls against their gross negligence and unmet responsibilities to animals—including the humblest creeping creatures, insects, and birds as well as useful livestock.

Nasr underscores the allegorical fabric of this and other epistles, in which animals, plants, minerals, and humans are microcosmic entities within the unity of creation, and at the same time hierarchically ordered within the great chain of being and harmonically related to one another.[42] Similar themes link nature, morality, and spiritual paths in the literatures of pilgrimage and Sufism.

In *The Conference of the Birds* by Farid ud-Din Attar, for example, birds with all different personalities and characters assemble to seek their king, the mythical *Simurgh*.[43] After many hesitations reflecting their diverse character weaknesses, thousands of birds set out on a most arduous journey to find their king. At last, thirty bedraggled birds survive to reach the *Simurgh* (lit. "thirty birds") with which they are all, despite their diversity, mystically united.

If these spiritual works seem far too far afield for modern geographers in the West, we may turn to al-Biruni's *India*, which is the most enduring work of cultural geography from the medieval Muslim world. By seeking to understand the values and geographic ideas of a foreign (Hindu) culture, al-Biruni speaks to the comparative and pluralistic concerns of this chapter.

Abu Raihan al-Biruni (973– c.1050 C.E.) was forcibly transported to Ghazna after the conquest of Khorezm in 1017 C.E. In addition to his many works on astronomy and mathematical geography, al-Biruni produced a massive study of Hindu India.[44] The purpose of his volume on India was to convey the "facts" and truth (*haqiqa*) about Hindu civilization, not to defend or refute it in the manner of Muslim or Hindu ideologues. In this task, he felt both alone among his Muslim colleagues and unwelcome among Hindus in whom he perceived a conceit against things and persons foreign or impure.

Al-Biruni began with the observation that Hindus seem to differ from Arab and Persian Muslims in every respect, including their language, religion, manners, and customs. His eighty chapters concentrate heavily on Hindu philosophy, literature, and sciences. They include geographic chapters on rivers and mountains, including brief comments on the "economy of nature."[45] Because his primary aim was to understand Hindu culture, however, al-Biruni was more concerned with Hindu concepts of nature than with putting forward his own description of the land, people, values, or natural environment. He noted the strong attachment to place among Hindus, to the territory bounded by the ecological limits of the gazelle, to pilgrimage routes and holy places including tanks, trees, and rivers.[46]

His Chapter 68 has special relevance for this chapter because it deals with what is allowed and forbidden in eating and drinking, including vegetarianism (2:202–5). Al-Biruni notes with some appreciation that Hindus, like early Christians and Manichaeans, forbade the killing and eating of meat, but he also asserts that, "People . . . have the desire for meat, and will always fling aside every order to the contrary" (2:202). He went on to sketch out several plausible explanations for the vegetarianism of Brahmans in India, including: (1) mytho-theological explanations derived from Hindu texts; (2) physiological explanations related to the heat of the plains and the "heat" of meats; and (3) economic explanations based on the value of animals for transport and agricultural labor compared with their value as food. He wrote, "I, for my part, am uncertain, and hesitate in the question of the origin of this custom" (2:204). He agreed with the Hindu view that only the ignorant require formal distinctions between things that are allowed and forbidden.

The significance of this chapter, and the work as a whole, lies in al-Biruni's sensitivity to alternative explanations, his attempts to objectively present them, his awareness of the limits of his understanding, his courage to quote the *Qur'an* alongside other religious texts, and his underlying goal of facilitating understanding and dialogue through comparative study. For example, he boldly compared the Hindu holy city of Benares with Muslim Mecca (2:197). He compared Hindu and Greek ideas because both civilizations were viewed as idolatrous by Muslims yet they had strong traditions of philosophy and mathematics, and al-Biruni believed there are similarities among the educated people of different cultures.

His commitment to constructive comparison was remarkable for eleventh century Afghanistan, and indeed for any era. But his work also raises questions of enduring relevance for geography and ethics. To what extent was al-Biruni's work made possible by the Muslim conquest of northwestern India? To what extent did it seek to facilitate further conquest through knowledge and dialogue with the learned elite? To what extent did it seek to rectify Muslim misperceptions and enhance appreciation and learning? These questions focus on the relations between knowledge and power, and they invite comparison with European and postcolonial geographic and environmental inquiry. Their continuing significance for Muslims in South Asia is evident in the millenary celebrations of al-Biruni's birth in 1973 in Pakistan.[47] In those celebrations, al-Biruni was honored as one of the greatest scientists, geographers, and comparativists of Muslim civilization. Scholars from around the world continue to study his attempts at fair, frank, incisive comparison.

Toward Comparison and Pluralism

The final task of this chapter is to consider how Muslim contributions to geography and ethics might speak to "live" issues and situations in the world today. What if anything endures of the inquiries of the Ikhwan al-Safa and al-Biruni? The Sincere Brethern believed that each new community, "takes over . . . all the sciences of the communities that have dominated previously."[48] From that perspective, pluralism is a transitional situation in which diverse traditions are subsumed within an emergent overarching philosophy. Al-Biruni, by contrast, sought to understand the similarities, differences, and ambiguities of a foreign culture. For him, pluralism was a situation to be scientifically and accurately understood. Just as ideas of classical Hinduism were remote for Muslims of the eleventh century, so too are Muslim geography and ethics unfamiliar to geographers and others in the United States today.

Geographers have explored the challenges of pluralism in other postcolonial contexts.[49] But these social perspectives on pluralism complicate the relations between geography and environmental ethics. On the one hand, they document the rich diversity of colonial, religious, and international environmental norms. On the other hand, they underscore the numerous fissures, incommensurables, and conflicts among those situations of "legal pluralism."[50] Geographic inquiry has only occasionally sought a constructive pluralism, as the recent literature on geography and imperialism clearly reveals.[51]

The problems of pluralism in Muslim societies vary from those of modernist states like Turkey to postcolonial states like Algeria, Egypt, India, and Pakistan, to orthodox Islamic states governed by *shari'a* law, like Saudi Arabia. Because religious ethics operate to some extent in each of these situations, the historical contributions of Muslim geography and philosophy have continuing relevance. Muslim contributions also have salience for transnational communities—expatriates, immigrants, students, travellers, and scholars—who regularly move between Muslim and non-Muslim places. Although environmental ethics may not be the primary concern of these groups, international comparison of environmental problems is becoming increasingly common.

Moreover, as multicultural and transnational situations proliferate, the need for cross-cultural normative comparisons will presumably grow. Unfortunately, normative geographic comparison remains underdeveloped in theory, method, and application.[52] Modern geographers seem poorly prepared to deal with environmental ethics rooted in different religious and social traditions. Despite renewed interest in the geography of religions, it retains a peripheral status in the discipline and in policy analysis.[53] However, in the Muslim contributions surveyed here, religious ethics are central to norms of resource use, and convictions about the nature, meaning, and value of the world and its creatures.

Geographers, by contrast, concentrate on political and economic explanations for environmental problems and, to a lesser extent, historical and behavioral explanations for environmental problems in different cultural contexts. Their disinterest in religious environmental ethics has several roots, including the well-worn debates about Lynn White Jr.'s thesis on Judeo-Christian impacts on the environment; general discomfort with religious discourse; and criticisms of religion as inherently mystifying, unscientific, and false.[54] Whatever the exact combination of reasons, it comes as little surprise that geographers have had little involvement in recent experiments to understand Muslim and other religious contributions to environmental ethics.

Under such circumstances, are geographic comparisons between Muslim and Euro-American environmental ethics possible, or useful? Under what circumstances might they lead toward, and not away from, a constructive pluralism? As a step toward answering these questions, and broadening the conversation within geography, we may consider five recent initiatives in international or comparative environmental ethics that have benefited from Muslim contributions and that might benefit from additional geographic perspectives and involvement.

1. *Dialogues between comparative philosophy and environmental ethics.* Interest in comparative environmental ethics dates to the mid-1980s when an environmental ethicist, Baird Callicott, and comparative philosopher, Roger Ames, collaborated to produce a collection of essays titled *Nature in Asian Traditions of Thought*. As a first effort, it raised important issues and generated healthy debates about the possibility of comparison and the (mis)uses of comparative inquiry. It acknowledged and sought to move beyond arguments that because

Asian philosophies had not prevented environmental degradation, little could be learned from them.[55] As Eugene Hargrove put it, "We are now ready I believe for a third period of study; one that treats environmental philosophy globally, while at the same time respects cultural diversity; one that goes beyond the broad categories of East and West and that takes full account of regional, national, and religious differences."[56]

As a first step, *Nature in Asian Traditions of Thought* exposed some basic problems, such as the continuing emphasis on "great" over "little" traditions; asymmetry in the treatment of religious ethics (alleged to be strong in Asia but superseded in the West); skepticism about the value of comparative philosophic inquiry, especially that which takes a pluralistic stance; and unfamiliarity with related fields of cross-cultural moral inquiry.[57] Regrettably, Islamic environmental ethics was not included in either Asian or Western traditions: one hopes that room will be found in future conferences of the East-West Philosophers' group.[58]

2. *Interfaith environmental dialogue*. Muslims have actively contributed to interfaith environmental discussions, in which they frequently draw upon the historical sources discussed in this chapter.[59] In a dialogue intended for a broad public audience, a spokesperson for Islam, Seyyed Hossein Nasr, focused on the destructive impact of modern Western social thought.[60] Although Nasr has demonstrated through his teaching, writing, and speaking in the United States a deep commitment to interfaith dialogue on the sacred, even his most appreciative colleagues wonder how much can be accomplished in pluralistic minority contexts such as the United States.[61] Institutions for interfaith dialogue with Muslims (e.g., The Duncan Black MacDonald Center at Hartford Seminary, which publishes *The Muslim World*; the Centre for the Study of Islam and Christian-Muslim Relations in England; and the Islam section of the Pontifical Council for Dialogue with other Religions at the Vatican) have yet to give sustained attention to environmental concerns.

3. *International environmental ethics*. Several efforts to bridge religious and secular ethics have focused on major international forums. One group of scholars has examined the ethical dimensions of AGENDA 21, the main strategy document of the United Nations Conference on Environment and Development at Rio de Janeiro in 1992. A Muslim contributor, Safei el-Deen Hamed provided a lucid synopsis of *shari'a* sources and principles of environmental management. He described the failings of colonial and secular ethics in Muslim

communities and concluded with the hope that "one could recognize a useful place for the *Shariah* in the total scheme of international development. . . . This does not mean that there should be a single ethic for all peoples, but that each society needs to develop and adopt a conservation ethic appropriate to its unique ecological context and in keeping with its particular cultural traditions."[62] Although these ideas (which invite comparisons with Montesquieu's *Spirit of the Laws*) were not as thoroughly discussed as issues of mainstream environmental ethics, several other panelists argued that religious ethics were essential for the implementation of AGENDA 21.

5. *International nongovernmental organizations*. Some of the most exciting international and cross-cultural experiments in recent decades, including those on environmental ethics, have involved nongovernmental organizations.[63] The International Union for the Conservation of Nature has convened the Ethics Working Group to draft a charter and covenant to complement its World Conservation Strategy. The Ethics Working Group has dealt extensively with relations between philosophical and religious environmental ethics.[64] It has published proceedings on the following topics: world ethics for sustainability, ethics and law, ethics and biodiversity, religion and environmental ethics, ethics and education, environment and development ethics, ecological citizenship, and international ethics organizations. To my knowledge, no geographers have been directly involved in the Ethics Working Group, and few are cited in its publications.

Several prominent Muslim lawyers and environmentalists are active in the Ethics Working Group. Significantly, they are contributing to international legal perspectives on environmental ethics, as well as explicitly Islamic concerns, and are thereby helping identify areas of common ground.[65] The group devoted considerable discussion to issues of universality and pluralism, and chose to seek common norms among different cultural and religious traditions of the world, in part to follow the powerful precedents set by the Geneva Accords and Universal Declaration of Human Rights.[66]

A more utopian, but at the same time activist, project is Elise Boulding's work on "building a global civic culture."[67] Boulding has worked with nongovernmental peace, development, and religious organizations, to envision or "image" ideal social relations and to launch practical initiatives to realize them. During the 1991 Persian Gulf War, she extended her project on imaging to Muslim organizations.

Another example of collaboration between Muslim and Western

nongovernmental organizations is titled "Pluralism and Its Cultural Expressions" (1992–93).[68] The project is jointly sponsored by the Aga Khan Trust for Culture and Rockefeller Foundation. Concerned about escalating contact and conflict among cultural groups in many parts of the world, the project focuses on practical activities that articulate and negotiate pluralistic situations in everyday life (vis-à-vis general theories of its desirability or possibility). The cultural expressions investigated to date include print journalism, cultural organizations, and cultural policies but not, as yet, grassroots environmentalism.

Conclusion

To my knowledge, geographers are not making a substantial contribution to these cross-cultural investigations of environmental ethics. Some explanations for why they have not, and arguments for why they could or should, have been presented in this chapter. Geographers are well placed by their study of different regions at various scales to shed light on cultural relations through comparative inquiry; to clarify the spaces, possibilities, and consequences of pluralism; and to foster constructive collaboration in different cultural and environmental contexts.

Although the contributions of medieval Muslim geographers and philosophers may seem less compelling than contemporary secular contributions, I have tried to show even how this seemingly arcane example may be salient in vast geographic areas and for many millions of people. But whichever peoples or places are involved, the first steps toward constructive pluralism are surely to recognize their existence and experience, to continually seek to understand their contributions, and to appraise their contemporary relevance in the scientific manner of al-Biruni, the mystical path of the *Ikhwan al-Safa,* and the formal religious and legal tradition of the *shari'a.* Only then may we be adequately prepared to assess the common ground, comparability, and prospects for constructive pluralism among the environments and peoples of the world.

Acknowledgments

This paper originated in a panel discussion entitled "Geography and Environmental Ethics: Overlapping Terrain," organized by Jim Proctor and Jody Emel at the 1995 meetings of the Association of American Geographers. I would like to thank Andrew Light and Jonathan Smith for encouraging its development into its present form, afterwhich it benefited from comments by Dr. Richa Nagar and graduate students in the history and theory of geography seminar at the University of Colorado, as well as referees for this publication. Remaining errors are mine.

Notes

1. In this chapter, the word "Islam" refers to religious traditions originating in revelations to the prophet Muhammad in the early seventh century C. E. (Common Era); "Muslim" refers to people who either profess those traditions or feel associated with them by lineage and culture.

2. Given its scope, this survey uses a very broad brush. While recognizing the pitfalls inherent in this approach, especially superficiality and overgeneralization, the survey approach still seems a useful introduction and overview. To offset these problems, the plurality of Muslim and Islamic traditions is stressed in the text and citations are offered to more detailed studies of specific topics. In addition, the chapter uses graphic diagrams, tables, and maps to help the reader visualize the broad-ranging set of relations, ideas, and fields of inquiry that are covered (for a rare but welcome graphic experiment by a philosopher in ethics, see L. C. Becker, "Places for Pluralism," *Ethics* 102 [1992]: 707–19). For further discussion of pluralism, see V. Held, "Moral Pluralism," in *Encyclopedia of Ethics,* ed. L. Becker and C. Becker, (New York: Garland Publishing, 1992), 2: 839–41; D. Wong, "Comparative Ethics," in *Encyclopedia of Ethics,* 1: 185–89.

3. Becker, "Places for Pluralism."

4. For some critical and constructive perspectives, see A. Godlewska and N. Smith, eds., *Geography and Empire* (Oxford: Blackwell, 1994); Hasan Uddin Khan, *Pluralism and its Cultural Expressions,* (three conference reports sponsored by The Rockefeller Foundation and the Aga Khan Trust for Culture, 1992-3, copies with author); J. L. Wescoat, Jr., "From the Gardens of the *Qur'an* to the Gardens of Lahore," *Landscape Research,* 20 (1995): 19–29; J. L. Wescoat, Jr., "The Right of Thirst for Animals in Islamic Law: A Comparative Approach," *Society and Space,* 13 (1996): 637–54; J. L. Wescoat, Jr., "Varieties of Geographic Comparison in *The Earth Transformed,*" *Annals of the Association of American Geographers* 84 (1994): 721–25; J. L. Wescoat, Jr., "Managing the Indus River Basin in Light of Global Climate Change: Four Conceptual Approaches," *Global Environmental Change: Human and Policy Dimensions* 1 (1991): 381–95.

5. Because this paper focuses on scholarly work, the term "West" refers to modern scholarly communities in the United States and Europe.

6. This paper draws upon experiences gained over ten years of studying water resources problems and landscape history in the Islamic Republic of Pakistan. The focus on "Islamic" and "Muslim" tradition in this paper reflects their pervasiveness in everyday discussions of social and environmental issues in Pakistan. In the aftermath of a flood, for example, I listened as people sought to sort out its human and divine dimensions and implications. I heard animal rights activists justify their interventions on behalf of abused of animals on religious grounds (J. L. Wescoat, Jr.,"The Right of Thirst for Animals in Islamic Law"). Interestingly, such arguments are not commonly applied by Pakistanis to issues deemed "environmental" in international usage today (e.g., air, water, and solid waste pollution); however, see A. Kader et al., *Basic Paper on the Islamic Principles for the Conservation of the Natural Environment* (Gland, Switzerland: IUCN and the Kingdom of Saudi Arabia, 1983).

7. M. W. Mikesell and A. B. Murphy, "A Framework for Comparative Study of Minority Group Aspirations," *Annals of the Association of American Geographers* 81 (1991): 581–604.

8. "Adherants of All Religions by Continental Area," in *The World Almanac* (Mahwah, N.J.: The World Almanac, 1995), 731.

9. Y. Y. Haddad, ed., *The Muslims of America* (New York: Oxford University Press, 1991); Y. Y. Haddad and Jane I. Smith, eds., *Muslim Communities in North America* (Albany: State University Press of New York, 1994); L. Poston, *Islamic Da'wah in the West: Muslim Missionary Activity and the Dynamics of Conversion to Islam* (Oxford: Oxford University Press, 1992).

10. C. L. Stone, "Estimate of Muslims Living in America, " in *The Muslims of America*, ed. Y. Y. (Oxford: Oxford University Press, 1991), 25–36.

11. These figures indicate the diversity of "Muslim" peoples and cultural traditions, which must be continually balanced against the common ground among Muslim groups and traditions. See Halliday for an insightful critique of essentialist and monolithic renderings of Islam on both sides of the "Orientalism" debate (Fred Halliday, " 'Orientalism' and its Critics," *British Journal of Middle Eastern Studies* 20 [1993]: 145–63).

12. J. B. Harley and D. Woodward, eds., *The History of Cartography*, vol. 2, book 1: *Cartography in the Traditional Islamic and South Asian Societies* (Chicago: University of Chicago Press, 1992); A. H. Siddiqui, "Muslim Geographic Thought and the Influence of Greek Philosophy," *Geojournal* 37 (1995): 9–16; J. L. Wescoat, Jr., "Resource Management: Oil Resources and the Persian Gulf Conflict," *Progress in Human Geography* 16 (1992): 243–56.

13. Joseph Esposito, ed., *The Oxford Encyclopedia of the Modern Islamic World*, 4 vols. (Oxford: Oxford University Press, 1994).

14. Although this approach is subject to the familiar critique of classicism in Western scholarship on Islamic cultures, I argue that it is offset by the variety and continuing vitality of the work that it identifies. This chapter includes nineteenth- and twentieth-century work by Muslim scholars and

others who have built upon the earlier works. There is a pressing need for continuing investigation, on a larger scale, of the transformations of Muslim geographic and environmental inquiry during Western expansion and subsequent postcolonial restructuring.

15. For example, see P. Crone, *Roman, Provincial, and Islamic Law: The Origins of the Islamic Patronate* (Cambridge: Cambridge University Press, 1987); B. Lewis, *Islam and the West* (Oxford: Oxford University Press, 1993); G. Makdisi, *The Rise of Humanism in Classical Islam and the Christian West* (Edinburgh: Edinburgh University Press, 1990); W. M. Watt, *Muslim-Christian Encounters: Perceptions and Misperceptions* (New York: Routledge, 1991).

16. For criticisms of academic "coverage" of Islam, see C. Bayly, "Knowing the Country: Empire and Information in India," *Modern Asian Studies* 27 (1993): 3 ff; E. Said, *Orientalism* (New York: Pantheon, 1978); E. Said, *Covering Islam: How the Media and the Experts Determine How We See the Rest of the World* (London: Routledge & Kegan Paul, 1981). For a counterpoint, see Halliday, " 'Orientalism' and its Critics."

Godlewska, Gregory and others break new ground on European geographical inquiry in Muslim countries, but they neglect the antecedents and reactions of Egyptians (A. Godlewska, "Map, Text and Image, The Mentality of Enlightened Conquerors: A New Look at the *Description de l'Egypte*," *Transactions of the Institute of British Geographers* 20 [1995]: 5–28; D. Gregory, "Between the Book and the Lamp: Imaginative Geographies of Egypt, 1849–50," *Transactions of the Institute of British Geographers* 20 [1995]: 29–57). The continuing importance of pre-colonial geographical perspectives and sources is evidenced in K. W. Butzer, "The Islamic Traditions of Agroecology: Cross-Cultural Experience, Ideas and Innovation," *Ecumene: A Journal of Environment, Culture, Meaning* 1 (1994): 7–50; H. Nast, "Islam, Gender, and Slavery in West Africa circa 1500: A Spatial Archaeology of the Kano Palace, Northern Nigeria," *Annals of the Association of American Geographers* 86 (1996): 44–77, and J. L. Wescoat, Jr., "Mughal Gardens and Geographic Sciences, Then and Now," *Muqarnas* (1996): forthcoming.

17. For geographic sources, see, for example, S. Maqbul Ahmed, "Cartography of al-Sharif al-Idrisi," in *Cartography in the Traditional Islamic and South Asian Societies*, ed. J. B. Harley and D. Woodward (Chicago: University of Chicago Press, 1992), 156–74; I. Habib, *The Atlas of the Mughal Empire: Political and Economic Maps with Detailed Notes, Bibliography and Index*, revised ed. (Delhi: Oxford University Press, 1990); A. T. Karamustafa, "Military, Administrative, and Scholarly Maps and Plans," in *Cartography in the Traditional Islamic and South Asian Societies*, pp. 209–27; F. Sezgin, *The Contribution of the Arabic-Islamic Geographers to the Formation of the World Map* (Frankfurt: Institut fur Geschicte der Arabish-Islamischen Wissenschaften, 1987).

18. For a discussion of the aims, strengths, and weaknesses of the "varieties of" genre in comparative geographic inquiry, see J. L. Wescoat, Jr., "Varieties of Geographic Comparison in *The Earth Transformed*."

19. For major reviews see Nasir Ahmad, *Muslim Contributions to Geography*, revised ed. (Lahore: M. Ashraf, 1965); S. Maqbul Ahmed, "Djughrafiya," in *Encyclopedia of Islam*, new ed. (Leiden: E.J. Brill, 1979); K. W. Butzer, "The Islamic Traditions of Agroecology"; A. T. Karamustafa, "Cosmographical Diagrams," in *Cartography in the Traditional Islamic and South Asian Societies*, 71–89; J. H. Kramers, "Djughrafiya," in *First Edition of the Encyclopedia of Islam, 1913–1936, Supplement*, vol. 9 (Leiden: E.J. Brill, 1979), 61–73; Siddiqui, "Muslim Geographic Thought and the Influence of Greek Philosophy"; G. R. Tibbetts, "Early Geographical Mapping," in *Cartography in the Traditional Islamic and South Asian Societies*, pp. 90–155.

20. Kramers, "Djughrafiya"; Ahmed, "Djughrafiya"; Tibbetts, "Early Geographical Mapping."

21. Karamustafa, "Military, Administrative, and Scholarly Maps and Plans"; J.M. Rogers, "Itineraries and Town Views in Ottoman Histories," in *Cartography in the Traditional Islamic and South Asian Societies*, pp. 228-55; Wescoat, "Mughal Gardens and Geographic Sciences, Then and Now."

22. S. Gole, *Indian Maps and Plans: From Earliest Times to the Advent of European Surveys* (Delhi: Manohar, 1989).

23. D. A. King and R. P. Lorch, "Qibla Charts, Qibla Maps, and Related Instruments," in *Cartography in the Traditional Islamic and South Asian Societies*, pp. 189–205.

24. Wescoat, "The Right of Thirst for Animals in Islamic Law"; Wescoat, "From the Gardens of the *Qur'an* to the Gardens of Lahore."

25. R. Walzer, "akhlak," in *Encyclopedia of Islam*, new ed. (Leiden: E.J. Brill, 1979), 1: 325–9.

26. M. Fakhry, 1991, *Ethical Theories in Islam* (Leiden: E.J. Brill, 1991); G. F. Hourani, *Reason and Tradition in Islamic Ethics* (Cambridge: Cambridge University Press, 1985); R. G. Hovannisian, ed., *Ethics in Islam* (Malibu: Undena Publications, 1985).

27. al-Ghazali, *The Faith and Practice of al-Ghazali*, trans. W.M. Watt (Oxford: Oneworld Publications, 1994), 39, 43.

28. Fakhry, *Ethical Theories in Islam;* S. H. Nasr, "Sacred Science and the Environmental Crisis: An Islamic Perspective," in *The Need for a Sacred Science* (Richmond, UK: Curzon Press, 1993), 129–148.

29. C.W. Troll, "A Note on an Early Topographical Work of Sayyid Ahmed Khan: *Asar al-Sanadid*," *Journal of the Royal Asiatic Society*, (1972): 135–44; C.W. Troll, *Sayyid Ahmed Khan: A Reinterpretation of Muslim Theology* (Karachi: Oxford University Press, 1979).

30. Allama Muhammad Iqbal, *The Reconstruction of Religious Thought in Islam*, 2d ed., ed. M. S. Sheikh (Lahore: Institute of Islamic Culture and Iqbal Academy of Pakistan, 1989 [1930]).

31. M.W. Dols, *Majnun: The Madman in the Medieval Islamic World*, trans. D.E. Immisch (Oxford: Clarendon Press, 1992), 337; Nizami, *Layla & Majnun*, trans. R. Gelpke (Boulder: Shambhala, 1978 [1966]).

32. E. Grube, "The Early Illustrated Kalilah wa Dimnah Manuscripts." *Marg* 43 (1991): 32–51; R. Wood, trans., *Kalila and Dimna: Selected Fables of Bidpai* (London: Granada, 1980).
33. Ahmed, "Cartography of al-Sharif al-Idrisi"; Kramers, "Djughrafiya."
34. Tibbetts, "Early Geographical Mapping."
35. Kramers, "Djughrafiya"; Tibbetts, "Early Geographical Mapping."
36. Y. Marquet, "Ikhwan al-Safa," in *Encyclopaedia of Islam,* vol. 3, new edition (Leiden: E.J. Brill, 1979), 1071–6; S.H. Nasr, *An Introduction to Islamic Cosmological Doctrines* (Boulder: Shambala, 1978).
37. Wood, trans., *Kalila and Dimna: Selected Fables of Bidpai*, 185–95.
38. Isma'il Ragi al-Faruqi, "On the Ethics of the Brethern of Purity," *The Muslim World* 50 (1960): 109–21, 193–8, 252–8; and 51 (1961): 18–24.
39. S. H. Nasr, "Sacred Science and the Environmental Crisis: An Islamic Perspective; Isma'il Ragi al-Faruqi, "On the Ethics of the Brethern of Purity."
40. L. Goodman, trans., *The Case of the Animals versus Man Before the King of the Jinn, A Tenth-Century Ecological Fable of the Pure Brethren of Basra* (Boston: Twayne Publishers, 1978).
41. Jinn are a diverse class of angelic beings.
42. Nasr, "Sacred Science and the Environmental Crisis: An Islamic Perspective."
43. Farid ud-Din Attar, *The Conference of the Birds, A Sufi Fable* (Boston: Shambala, 1993).
44. al-Biruni, *Al-Beruni's India,* trans. E. Sachau (Lahore: Sheikh Mubarak Ali, 1963 [reprint]); see also, D. J. Boilot, "al-Biruni," in *Encyclopedia of Islam,* new edition (Leiden: E.J. Brill, 1979), 1: 1236–8; Edward S. Kennedy, *A Commentary on Biruni's Kitab tahdid al-amakin*; an 11th Century Treatise on Mathematical Geography (Beirut: American University of Beirut, 1973); al-Biruni, *Athar al-baqiyah. The Chronology of Ancient Nations,* trans. and ed. by C. E. Sachau (London: W.H. Allen, 1879).
45. Nasr, "Sacred Science and the Environmental Crisis: An Islamic Perspective."
46. al-Biruni, *Al-Beruni's India,* 2: 191. See also, M. Marriott, "On 'Constructing an Indian Ethnosociology,' " *Contributions to Indian Sociology* (n.s.) 25 (1991): 295–308; Francis Zimmerman, *The Jungle and the Aroma of Meats: An Ecological Theme in Hindu Medicine* (Berkeley: University of California Press, 1993).
47. Hakim Muhammad Said, *Al-Biruni Commemorative Volume: Proceedings of the International Congress Held in Pakistan, November 26th thru December 12th, 1973* (Karachi: Hamdard Academy, 1979).
48. Marquet, "Ikhwan al-Safa," 1073.
49. C. Clarke, D. Ley, and C. Peach, eds., *Geography and Ethnic Pluralism* (London: George Allen & Unwin, 1991); Mikesell and Murphy, "A Framework for Comparative Study of Minority Group Aspirations."
50. M. Hooker, *Legal Pluralism: An Introduction to Colonial and Neo-*

Colonial Laws (Oxford: Clarendon Press, 1975); A. Watson, *Legal Transplants: An Approach to Comparative Law* (Charlottesville: University Press of Virginia, 1974); A. Watson, "Legal transplants and law reform," in *Legal Origins and Legal Change* (London: The Hambledon Press, 1991), pp. 293–8.

51. Godlewska and Smith, eds., *Geography and Empire*.

52. Wescoat, "Varieties of Geographic Comparison in *The Earth Transformed.*"

53. A. Buttimer, "Geography, Humanism, and Global Concern," *Annals of the Association of American Geographers* 80 (1990):1 ff; C. Park, *Sacred Worlds: An Introduction to Geography and Religion* (London: Routledge, 1994).

54. On the White thesis, see J. B. Callicott and R.T. Ames, eds, *Nature in Asian Traditions of Thought: Essays in Environmental Philosophy* (Albany: SUNY, 1989); J. Kay, "Human Dominion over Nature in the Hebrew Bible," *Annals of the Association of American Geographers* 79 (1989): 214 ff; Yi Fu Tuan, "Discrepancies between Environmental Attitude and Behavior: Examples from Europe and China," *Canadian Geographer* 12 (1968): 176-91. On the general discomfort with religious discourse, see J. A. Berling, "Is Conversation about Religion Possible?" *Journal of the American Academy of Religion* 61 (1992): 1–22.

55. Tuan, "Discrepancies between Environmental Attitude and Behavior."

56. In Callicott and Ames, eds., *Nature in Asian Traditions of Thought*, xxi.

57. R. A. Shweder, M. Mahapatra, and J. G. Miller, "Culture and Moral Development," in *Cultural Psychology: Essays on Comparative Human Development* (Cambridge: Cambridge University Press, 1990); Michael Stocker, *Plural and Conflicting Values* (Oxford: Clarendon Press, 1990).

58. *Philosophy East and West*, special issue: The Sixth East-West Philosopher's Conference, "Culture and Modernity: The Authority of the Past," 41 (Oct 1991); Wong, "Comparative Ethics."

59. K. L. Afrasiabi, "Towards an Islamic Ecotheology," *Hamdard Islamicus: Quarterly of the Hamdard National Foundation, Pakistan* 18 (1995): 33–50; M. Deen, "Islamic Environmental Ethics, Law and Society," in *Ethics of Environment and Development: Global Challenge, International Response*, ed. J. Engel and J. Engel (Tucson: University of Arizona Press, 1990), 189–98; Nasr, *An Introduction to Islamic Cosmological Doctrines;* S. H. Nasr, "The Ecological Problem in the Light of Sufism: The Conquest of Nature and the Teachings of Eastern Science," in *Sufi Essays*, 2d ed. (Albany: State University of New York, 1991), 152–63; Nasr, "Sacred Science and the Environmental Crisis."

60. S. C. Rockefeller and J. C. Elder, *Spirit and Nature: Why the Environment is a Religious Issue* (Boston: Beacon Press, 1992).

61. See issues of *Journal, Institute of Muslim Minority Affairs*. See also, G. Kramer, "Minorities: Minorities in Muslim Societies," in *The Oxford Encyclopedia of the Modern Islamic World*, (Oxford: Oxford University Press,

1995), 3: 108–10; J. I. Smith, "Seyyed Hossain Nasr: Defender of the Sacred and Islamic Traditionalism," in *The Muslims of America*, 80–95.

62. D. A. Brown, ed., "Proceedings . . . The Ethical Dimensions of the United Nations Program on Environment and Development, AGENDA 21," (Photocopy, 1994), 109.

63. Although I focus in this survey on large international nongovernmental organizations, more fine-grained research is warranted on Muslim NGO's, grassroots organizations, and ethnographies. For an insightful study by a Western geographer on environmental work and education in a Muslim village, see C. Katz, "Sow What You Know: The Struggle for Social Reproduction in Rural Sudan," *Annals of the Association of American Geographers* 81 (1991): 488–514.

64. J. H. Callewaert, "International Documents and the Movement toward a Global Environmental Ethic," (Photocopy with author, 1994); J. R. Engel and J. Denny-Hughes, *Advancing Ethics for Living Sustainably*, Report of the IUCN Ethics Workshop (Sacramento: International Center for Environment and Public Policy, April 1993); IUCN, "Papers from the Ethics and Covenant Workshop, IUCN General Assembly," (Buenos Aires, Argentina, January 1994).

65. M. Helmy, *Islam and Environment 2—Animal Life* (Kuwait City: Environmental Protection Council, 1989); Kader et al., *Basic Paper on the Islamic Principles for the Conservation of the Natural Environment;* A. Masri, *Islamic Concern for Animals* (Petersfield: The Athene Trust, n.d); A. Masri, "Animal experimentation: The Muslim viewpoint," in *Animal Sacrifices: Religious Perspectives on the Use of Animals in Science,* ed. T Regan (Philadelphia: Temple University Press, 1986), 171–98; J. O'Brien and F. Khalid, eds., *Islam and ecology* (New York: Cassell, 1992).

66. R. Afshari, "An Essay on Islamic Cultural Relativism in the Discourse on Human Rights," *Human Rights Quarterly* 16 (1994): 235–76; A. Naim, ed., *Human Rights in Cross-Cultural Perspectives* (Philadelphia: University of Pennsylvania Press, 1992); B. Tibi, "Islamic Law/*shari'a*, Human Rights, Universal Morality, and International Relations" *Human Rights Quarterly* 16 (1994): 277–99.

67. E. Boulding, *Building a Global Civic Culture: Education for an Interdependent World* (Syracuse: Syracuse University Press, 1988).

68. I am grateful to Hasan Uddin Khan, formerly of the Aga Khan Trust, for information about this project.

The Dialectical Social Geography of Elisée Reclus

John Clark

While Elisée Reclus is still recognized as an important figure in both the history of geography and the history of anarchist political theory, his thought has been given little careful examination in recent times.[1] This is unfortunate, since his ideas are even more relevant today than they were in his own day, when he was widely known as the foremost geographer of France, and feared by many as a dangerous political radical. Indeed, a careful study of his thought shows him to be not only a pioneering figure in social geography, but also an ecological social theorist who long ago explored areas that have become central concerns of environmental philosophy and environmental ethics today. Perhaps most notably, Reclus is an important precursor of social ecology, which is widely considered to be one of the three major tendencies in contemporary radical ecological theory.[2] This paper will focus on one important aspect of his impressive body of thought: his holistic, dialectical interpretation of the place of humanity in the natural world. First, however, a brief discussion of Reclus's place in the political and intellectual life of his time may be helpful in putting his ideas in historical context.

I

Reclus's career as a pioneering and impressively prolific geographer spans over half a century. Beginning in the 1860s, he began publishing articles in the *Revue des deux mondes* and many other journals, and completed the first of his three great geographical projects, *La Terre: description des phénomènes de la vie du globe*.[3] Its two volumes, running to over fifteen hundred pages, were published in 1867 and 1868. Though still in his thirties at this time, Reclus was already gaining wide recognition as an important geographer.

Reclus's intellectual work was interrupted in the early 1870s by the events of the Paris Commune and its aftermath. He personally participated both in the politics of the Commune and in the defense of Paris. His column of the Paris National Guard was taken prisoner by the victorious Versailles troops and he spent the next eleven months in fourteen different prisons. He was later tried and sentenced to deportation in New Caledonia, but because of his prestige as a scientist and intellectual, his friends and supporters succeeded in having his sentence reduced to ten years exile. As a result, he was allowed to emigrate to Switzerland, where he began his association with the anarchists of the Jura Federation, and developed close ties with the major anarchist theorists Mikhail Bakunin and Peter Kropotkin.

It was also in Switzerland that he began his greatest work, the *Nouvelle géographie universelle*.[4] This monumental achievement, which ran to seventeen thousand pages, appeared in nineteen volumes between 1876 and 1894. According to geographer Gary Dunbar, in his biography of Reclus, "for a generation the *NGU* was to serve as the ultimate geographical authority" and constituted "probably the greatest individual writing feat in the history of geography."[5] Reclus remained in Switzerland until 1890, heavily occupied with both scholarship and political activity, and then finally returned to France.

In 1894, Reclus began a new phase of his career when he accepted an invitation to become a professor at the New University in Brussels. He had some reservations about this undertaking, having remained outside the academic world until quite late in life. However, he was a great success in this role, achieving renown as a teacher and winning the enduring admiration of many students. During this period he also completed his last great work, *L'Homme et la Terre*, which he completed shortly before his death in 1905.[6] This impressive, wide-ranging study in six volumes and thirty-five hundred pages reinforced his reputation as a major figure in the history of geography. It is in this final work that Reclus's most extensive and most sophisticated discussions in social theory are to be found.

While Reclus's social geography makes an important contribution in many areas of scholarship, his most enduring intellectual legacy is his contribution to the development of an ecological worldview, and to ecological social thought, in particular.[7] Béatrice Giblin contends that Reclus "had a global ecological sensibility that died with him for almost a full half-century."[8] This sweeping generalization is perhaps even an understatement. The kind of ecological perspective that Reclus developed, especially in his *magnum opus* of social theory, *L'Homme et la Terre*, effectively disappeared from social thought for most of the century, and did not reemerge into the intellectual mainstream until well into the 1970s, in response to growing public awareness of the ecological crisis. And, indeed, his sweeping account of humanity's integral development within a larger earth history has been unparalleled until Thomas Berry and Brian Swimme's *The Universe Story* was published in 1992.[9]

II

Reclus begins the first volume of *L'Homme et la Terre* with the epigraph: "Man is nature becoming self-conscious."[10] This proposition (which in the original French states literally that humanity is "nature taking consciousness of itself") captures the essence of Reclus's message: that humanity must come to understand its identity as the self-consciousness of the earth, and that it must in its own historical development realize the profound implications of this identity. In effect, Reclus proposes to humanity an ethical project of taking full responsibility, through a transformed social practice, for our place in nature, and a corresponding theoretical project of more adequately understanding that place and of unmasking the ideologies that distort it. Accordingly, he seeks to explain the development of human society in its dialectical interaction with the rest of the natural world, and expounds a theory of social progress in which human self-realization and the flourishing of the planet as a whole can be reconciled with one another. In these goals, Reclus's problematic intersects with the most central concerns of recent ecological thought.

Reclus exhibits in all his works a strong sense of humanity's embeddedness in nature. Even in his early work, he eloquently describes humanity as an expression of the earth's creativity and stresses human kinship with the entire system of life. "We are," he says, "the children of the 'beneficent mother,' like the trees of the forest and the reeds of the rivers. She it is from whom we derive our substance; she nourishes us with her mother's milk, she furnishes air to our lungs, and, in fact, supplies us with that wherein we live and move and have our being."[11] Throughout his works, he continues to develop this holistic, integrative outlook. While over the course of his career his studies of the natural world became increasingly scientific and empirical, he never abandoned his early romanticist, poetic, moral, and spiritual attitudes toward nature. Indeed, his resultant effort to integrate forms of rationality with aesthetic and moral sensibility (in effect, to unite the quest for the true, the beautiful, and the good) is one of the most noteworthy dimensions of his thought.

One aspect of this endeavor is his effort to synthesize a theoretical and scientific understanding of nature with an awareness of the practical implications of such an understanding. The result can be seen as a kind of politics of self-conscious nature, a thoroughly political geography that anticipates today's political ecology. Yves Lacoste, one of the contemporary French geographers who has done the most to revive interest in Reclus, contends that while Reclus was "the greatest French geographer," he has been "completely misunderstood" because of the "central epistemological problem of academic geography: the exclusion of the political."[12] Lacoste finds it ironic that recent discussions of social geography systematically "forget" Reclus's massive six-volume work in which social geography is itself the "main thread."[13]

The surprisingly far-reaching conception of social geography found in *L'Homme et la Terre* contributed much to the development of a dialectical, holistic view of nature. For example, Reclus accepts the dialectical principle that every phenomenon embodies in itself the entire history of that phenomenon. He utilizes this principle when he observes that "present-day society contains within itself all past societies," and applies it to human nature in general, adopting a version of the doctrine that "ontogeny recapitulates phylogeny."[14] In his variant of this theory, "Man recollects [*remémore*] in his structure everything that his ancestors lived through during the vast expanse of ages. He indeed epitomizes [*résume*] in himself all that preceded him in existence, just as, in his embryonic life, he presents successively various forms of organization that are more simple than his own."[15]

In accord with this dialectical approach, Reclus believes that an examination of the history of the evolution of human society can guide us in understanding the structure and contradictions of present-day society. In his analysis of modern societies, Reclus discovers that each of them "is comprised of superimposed classes, representing in this century all successive previous centuries with their corresponding intellectual and moral cultures," and that when they are "seen in close juxtaposition, their vastly differing conditions of life present a striking contrast."[16] In his investigation of these classes, Reclus seeks to uncover certain fissures in the social structure that are usually concealed under layers of ideological mystification. It can thus be shown how the hidden legacy of social domination reveals itself in contemporary social conflicts.

Reclus holds that in order to transcend that legacy, humanity must develop a critical consciousness of past historical development. Such awareness can offer a basis for consciously creating a future collective history. He describes this process as humanity's attempt "to realize itself through one form that encompasses all ages."[17] As the species comes to see itself as part of a historical and geographical whole (and thus, a temporal and spatial one), it gains both self-consciousness and a corresponding freedom. We achieve the ability "to free ourselves from the strict line of development determined by the environment that we inhabit and by the specific lineage of our race. Before us lies the infinite network of parallel, diverging, and intersecting roads that other segments of humanity have followed."[18]

While an "ecological" perspective was once identified with a one-sided emphasis on harmony, balance, and order, recent discussions in ecological theory have challenged the classical "ecosystems" model. In fact, some theorists, inspired by postmodernist thought, have embraced the opposite extreme, seeing only disorder and chaos in nature. Reclus long ago supported a much wiser dialectical view that avoids both the static and chaotic extremes.[19] There is indeed, according to Reclus, a harmony and balance in nature, but it is one that operates through a tendency toward discord and imbalance. He notes that "as plants or animals, including humans, leave their native habitat and intrude on another environment, the harmony of nature is temporarily disturbed"; however, these introduced types either die out or adapt to the new conditions, making a contribution to nature as they "add to the wonderful harmony of the earth, and of all that springs up and grows upon its surface."[20] The balance of nature is thus a balance of order and disorder.

Reclus's strongly holistic view of nature often sounds strikingly similar to contemporary ecological analyses. An example is his discussion of the function of forests in global ecological health. He laments the reckless and destructive actions of the "pioneers" of both North and South America, who burned huge expanses of ancient forest in order to establish agriculture, "at the same time burning the animals, blackening the sky with smoke, and casting to the wind ashes that scatter over hundreds of kilometers."[21] He notes that while this action was shortsighted even from an economic point of view, the great loss is that the forests have been prevented from playing "their part in the general hygiene of the earth and its species," which is "an essential role."[22] Reclus uses strongly organicist imagery to present a model of ecological soundness conceptualized as health, and he shows the links between human health and ecosystemic health. The earth, he says, "ought to be cared for like a great body, in which the breathing carried out by means of the forests regulates itself according to a scientific method; it has its lungs which ought to be respected by humans, since their own hygiene depends on them."[23] He also uses aesthetic images to express this same holistic, organicist view of nature, as when he describes the earth as "rhythm and beauty expressed in a harmonious whole."[24]

One of the most widely debated concepts in recent ecological thought is "anthropocentrism," which is often defined as an outlook that places human beings in a hierarchical position over all other beings, and which reduces all value in "external nature" to a merely instrumental one in relation to human ends. Reclus sometimes uses language that sounds distinctly "anthropocentric," as when he writes of the "conquests" involved in human progress. However, the major import of Reclus's social geography is to remove humanity from a position above or over against the natural world, and to incorporate it fully into the life and history of the planet. What is striking about Reclus's viewpoint is the degree to which he could, unlike so many other nineteenth-century thinkers, shift from a human-centered to an earth-centered perspective.

Rather than being "anthropocentric," Reclus's view of the place of humanity in nature centers around the larger whole of nature of which we are a part, and the larger processes of development in which we participate. In a sense, Reclus's view may be called an "emergence" theory, if it is understood that he sees humanity as emerging *within* nature rather than *out of* it. His analysis prefigures in some ways Murray Bookchin's division of the natural world into a "first nature" and a "second nature," corresponding more or less to the natural world and the social world, both of which are seen as developing forms of "nature."[25] Reclus delineates similar realms of being within the natural world. There is, on the one hand, that sphere of nature that exists independent of humanity, and that had, indeed, existed for aeons before nature began to "become conscious of itself" through the development of humanity. As humanity emerges, it remains in intimate interrelationship with an external sphere of nature, and the complex relationships of interdependence between the two realms take on an increasingly planetary dimension. Reclus calls the realm of natural being that has arisen and related itself to the rest of nature "the human social milieu."

However, the human social world does not constitute for Reclus a single "second nature," for it is itself dual, and might be said to encompass both a "second nature" and a "third nature." He calls the former "the static milieu" or "the natural conditions of life," and he labels the latter "the dynamic milieu" or "the artificial sphere of existence." The former sphere, even though it is shaped, in a sense, by human culture, constitutes our most immediate embeddedness in nature, and thus still remains in some ways a realm of natural necessity. The latter sphere is much more subject to human direction and is much more profoundly shaped by social contingency. For Reclus, there is

> a quite marked distinction between the facts of nature, which are impossible to avoid, and those which belong to an artificial world, and which one can flee or perhaps even completely ignore. The soil, the climate, the type of labor and diet, relations of kinship and marriage, the mode of grouping together, these are the primordial facts that play a part in the history of each man, as well as of each animal. However, wages, ownership, commerce, and the limits of the state are secondary facts.[26]

Reclus's discussion here does not seem entirely coherent. On the one hand, some of his "facts of nature" seem eminently cultural, as in the case of kinship systems. On the other hand, while systems of commerce are profoundly cultural, they are also an expression of the quite "natural" needs to produce and to exchange products in some way. However, Reclus still seems to be making an important point. While all human activity is cultural, there seem to be certain "facts of nature" that *require* a cultural expression, and there are certain "facts of culture" that seem to be relatively autonomous from natural necessity. In defense of the arbitrariness of the institutions he associates with "secondary facts," he observes that many earlier societies managed to exist without them. He argues for the theoretical priority of the "static milieu," since it has always existed, and has often had a determining force in social affairs. Although he admits that "quite often in the case of individuals the artificial sphere of existence prevails over the natural conditions of life," he thinks that "it is necessary to study the static milieu first and then to inquire into the dynamic milieu."[27] This statement does not seem particularly dialectical, since the important question is not which sphere is considered first, but rather whether the mutual determinations between them are investigated adequately.

But when he considers the relationship between the two spheres, he does see it as a dialectical one. He is particularly concerned that the place of nature in the dialectic should be given adequate attention. Reclus contends that the influence of nature and of the "static milieu" in determining the character of social phenomena is much greater than historians and social theorists have previously recognized. He states that in the development of society over history, "nothing is lost," for "the ancient causes, however attenuated, still act in a secondary manner, and the researcher can discover them in the hidden currents of the contemporary movement of society."[28] While superimposed political and economic factors are often given primary recognition as social causes, "this second dynamic milieu, added to the primitive static milieu, constitutes a whole of influences within which it is difficult, and often even impossible to determine the preponderance of forces. This is all the more true because the relative importance of primary and secondary forces, whether purely geographical or already historical, varies according to peoples and ages."[29] Once again, the phenomenon—including even the social whole—can only be understood as the cumulative product of its entire history. Indeed, humanity itself, "with all its characteristics of stature, proportion, traits, cerebral capacity" is "the product of previous milieux multiplying themselves to infinity" since the origins of the species.[30]

Reclus may be seen as a precursor of bioregional thinking, in so far as he concludes that we and our cultures reflect the earth and the specific regions of the planet in which we have developed. In his words, "the history of the development of mankind has been written beforehand in sublime lettering on the plains, valleys, and coasts of our continents."[31] While bioregionalism has only recently become an important tendency in ecological thought, Reclus long ago recognized that we are, in our very being, regional creatures.[32] Yet, as is the case for every relation, that existing between humanity and the earth and its regions is also a dialectical one. It results from mutual interaction, as the earth expresses itself through humanity, and as humanity acts upon the earth. And Reclus recognizes that this interaction includes humanity's struggle with the rest of the natural world. Thus, "the accordance which exists between the globe and its inhabitants" cannot be described adequately through a one-sided focus on terms like "harmony," "balance," and "oneness" that exaggerate the degree to which order prevails, since whatever order that exists "proceeds from conflict as much as from concord."[33] The interrelationship between humanity and the earth is a process of dynamic mutual determination.

Reclus is especially interested in analyzing the side of this interrelationship that has been neglected by much of social thought throughout the modern period: the conditioning of the "social" by the "natural." His position on this subject should not be confused with the tradition that begins with Montesquieu's famous speculations on the influence of climate on society.[34] In such discussions, the appeal to natural influences becomes little more than an attempt to give an "objective" basis to the writer's social and cultural prejudices, so that characteristics attributed to various peoples become essential qualities that dictate strict limits for possible social change. This tradition culminates in theories such as Ellsworth Huntington's "human geography," in which the appeal to nature becomes the ideological justification for white supremacy and European hegemony.[35] Reclus's analysis should be distinguished from such views not only on the basis of his differing value-commitments, but also by his radically different methodology. He is interested in a dialectic between nature and culture, and on the interaction between a great many natural and social factors that shape human society. Far from attributing rigidly determined, almost immutable qualities to peoples and cultures, he hopes that by understanding the determinants of the social world, all peoples can ultimately become active, conscious agents in their own liberation. His analysis helps remind us that the investigation of the influence of the natural world on cultural practices and social institutions does not necessarily have reactionary implications.

Reclus offers the history of ancient religions as an example of the influence of natural geography on social institutions. He suggests that the monotheism of the ancient Near East reflects the austere character of that region's terrain. He remarks that one might generalize "that throughout the Semitic countries the splendid uniformity of tranquil spaces, illuminated by a violent sunlight, must have contributed mightily to giving a noble and serious turn to the concepts of the inhabitants. They learned to see things simply, without searching for great complications."[36] He contrasts this unifying vision to the unity-in-diversity expressed in Indian religion. The Near Eastern mythology "bore no resemblance to the chaos of divine forces leaping out of nature in infinite variation that one finds in India, with its high mountains, great rivers, immense forests, and climate whipped into rages by the abundant rains and the fury of storms."[37] Reclus notes that the "Hindu spirit" also perceived an underlying order and unity in the cosmos, but it naturally expressed this "single force" in "an infinite variety" of manifestations.[38]

Reclus does not, it should be stressed, attempt to reduce the complexity of religious phenomena (or any others) to a mere reflection of geographical qualities. Indeed, he often puts at least as much emphasis on the significance of the economic, the technical, and other "material" determinants, not to mention the political ones, in shaping all aspects of society. But in an age in which other determinants (and, specifically—under the influence of capitalist, socialist, and even some anarchist ideology—the economic and technological ones) were attributed enormous significance, he wished to emphasize the general neglect of the influence of the natural world on human history. His philosophy is noteworthy for the degree to which it uncovers, beneath the historical dialectic of institutions and experience, an active natural world that continually exercises its influence through certain geographical factors that are often overlooked in our focus on human transformative activity.

Reclus emphasizes the need for a greater recognition of nature, not only in the sense of understanding its activity, but also in the sense of developing a new responsibility toward it. This concern underlies the scathing critique of humanity's abuse of the earth that he began to develop early in his work. In "The Sense of Nature," he writes of the "secret harmony" that exists between the earth and humanity, and warns that when "reckless societies allow themselves to meddle with that which creates the beauty of their domain, they always end up regretting it."[39] When humanity degrades the natural world, he concludes, it thereby degrades itself. Reclus's analysis of this phenomenon is very close to the view recently developed by Thomas Berry, who argues that the diversity and complexity of the human mind reflects the richness and complexity of the earth and its regions, so that in damaging the earth, we harm ourselves not only physically, but in our "intellectual understanding, aesthetic expression, and spiritual development."[40] Reclus states similarly that "where the land has been defaced, where all poetry has disappeared from the countryside, the imagination is extinguished, the mind becomes impoverished, and routine and servility seize the soul, inclining it toward torpor and death."[41] And he does not neglect the material damage to human society caused by ecological degradation. He notes that "the brutal violence with which most nations have treated the nourishing earth" has been "foremost among the causes which have vanquished so many successive civilizations."[42]

In accord with his general view that the good and the beautiful tend to accompany one another, he links our ethical obligations to the natural world with our aesthetic appreciation of it. He gives an example of this link in the case of the domestication of animals, which he considers an intolerable abuse of nature. He notes not only the callousness with which animals are treated, but the "hideousness" of the results of this process, in which animals bred for human purposes lose both their adaptive qualities and their natural beauty.[43] And just as he links the act of harming nature to the creation of ugliness, he associates acting in accord with the good of nature with the creation of beauty. "Man," he says, can find beauty in "the intimate and deeply-seated harmony of his work with that of nature."[44] But our obligations in this sphere go beyond this complementary activity. Beyond acting in harmony with nature, we must engage ourselves also in the active defense of it. In view of the fact that "a reckless system has defaced that beauty," it is necessary for "man" to "endeavor to restore it" through efforts to "repair the injuries committed by his predecessors."[45]

In Reclus's holistic conception of humanity-in-nature, humanity's striving to achieve beauty and harmony should be seen as an integral part of the creation of these qualities throughout the natural world. Thus, "man" should "assist the soil instead of inveterately forcing it," in order to achieve "the beautification as well as the improvement of his domain," by giving "an additional grace and majesty to the scenery which is most charming."[46] Human creative self-expression will thus cooperate with the larger processes of creative self-expression in nature. Our goal in life should be "making our existence as beautiful as possible, and in harmony, so far as we are capable, with the aesthetic conditions of our surroundings."[47] For Reclus there is a continuity between our concern for ourselves, for others, and for the earth. "Ugliness in persons, in deeds, in life, in surrounding nature— this is our worst foe. Let us become beautiful ourselves, and let our li[ves] be beautiful!"[48] According to Reclus's holistic conception of human nature, as "man" becomes aware of the implications of being "nature becoming self-conscious," and thus "the conscience of the earth," "he" will naturally accept "responsibility as regards the harmony and beauty of nature around him."[49]

Reclus recognizes, of course, that we are far from achieving such a harmonious and cooperative relationship with the earth. He laments the fact that we become so engrossed in the process of transforming nature through labor according to our narrow technical and economic ideas that we fail to recognize nature's own creative powers. He urges us to learn to appreciate the integrity of the earth, so that we may cooperate with it in achieving various goods, instead of seeking to impose our will on it. In his view, "when man forms some loftier ideal as regards his action on the earth, he always perfectly succeeds in improving its surface, although he allows the scenery to retain its natural beauty."[50] Agriculture, for example, must not be reduced to a process of mining the soil of its nutrients for the sake of productivity. It is necessary, instead, to "comprehend" the land and to "humor" it by discovering which crops suit it best. In a recognition of the importance of imagination, sensibility, and symbolic expression, he praises the Shakers for mutualistic practices that make agriculture a "ceremony of love" in which all aspects of nature are "cherished."[51]

Much like Kropotkin, his fellow anarchist geographer, Reclus looks for contemporary models for a more balanced and humane relationship to the natural world. He notes certain examples in Europe of the way in which agricultural productivity can be reconciled with the beauty of the landscape. Writing in the 1860s, he remarks that "a complete alliance of the beautiful and the useful" has been attained in certain areas of England, Lombardy, and Switzerland, places where agriculture is in fact "most advanced."[52] He also cites as instances of such a beneficial alliance the draining of marshes in Flanders to produce farmland, the irrigation of the barren Crau region, the planting of olive trees along the slopes of the Apennines and Alps, and the replacement of Irish peat bogs by diverse forests.[53]

While these examples may support Reclus's contention that humans can contribute to beauty in nature, they also show certain flaws in his outlook from an ecological point of view. Although the kind of projects he cites have sometimes increased natural beauty, his examples show his bias toward "humanized" landscapes. He seems less sensitive to the natural beauty of, for example, the more austere terrain of rugged mountains, or the rich wildness of a swampland. Similar criticisms have sometimes been directed toward Bookchin's version of social ecology. In both cases, however, the writers' pastoralist emphasis reflects the way in which their own proclivities conditioned their versions of these theories, rather than any fundamental limitation of the applicability of either social geography or social ecology. Both theories are based on a dialectical view of the relationship between humanity and nature, a holistic analysis of phenomena that stresses the importance of unity-in-diversity, and a commitment to nondomination and spontaneous development. These theories are therefore fully capable of grasping the place of wilderness and "free nature" in the processes of natural unfolding.[54]

This is not to minimize the self-contradictory nature of certain aspects of Reclus's thought (or Bookchin's, for that matter). In Reclus's works, one finds an implicit contradiction between his developing holistic, ecological perspective and remnants of the dualistic, human-centered outlook that was so common in his age. In an early work, he exhibits the latter tendency strongly when he remarks favorably that science is "gradually converting the globe into one great organism always at work for the benefit of mankind."[55] This rather extravagant conception of the earth's processes as a vast conspiracy to benefit our species is far from his later, more developed holistic perspective. There, humanity is integrated into the planetary whole as the consciousness of the earth, and the healthy functioning of the earth's metabolism benefits humanity only as one part of that flourishing whole. In the passage just cited, Reclus says that human transformative activity has the capacity to make the earth into "that pleasant garden which has been dreamed of by poets in all ages."[56] Such an image expresses Reclus's enduring ideal of a harmonious relationship between humanity and nature, but errs in the direction of stasis, omitting the element of dialectical tension that must always characterize human confrontation with the otherness of nature.[57] Further, it can easily be taken to imply the desirability of the destruction of the wildness and freedom of the natural world, and to idealize a domesticated, highly humanized nature that is far from being an authentically ecological conception. Such themes become more muted in Reclus's later works, but they do not disappear entirely.[58]

On the other hand, Reclus was from the outset a forceful critic of the more blatant forms of human destructiveness toward nature that were accepted with complacency by many of his contemporaries. He judges that in civilization's dealings with nature, "everything has been mismanaged," so that what is left is "a pseudo-nature spoilt by a thousand details—ugly constructions, trees lopped and twisted, footpaths brutally cut through woods and forests."[59] Like later social ecologists, he sees the problem as both ideological and institutional. Looking at its subjective dimension, he points out that human interaction with nature has not been guided by "a sentiment of respect and feeling" for nature, but rather by "purely industrial or mercantile interests."[60] For this to change, a revolution in values must certainly take place. But this ideological transformation can only succeed if there is a complementary process of social transformation. An attitude of "respect and feeling" can prevail only if the social order based on disrespectful and unfeeling interests—for example, "industrial or mercantile" ones—can be eliminated. The ultimate union between "the civilized" and the "savage" and between humanity and nature can take place "only through the destruction of the boundaries between castes, as well as between peoples."[61] This implies for Reclus the abolition of the system of economic inequality embodied in capitalism, the system of political domination inherent in the modern state, the system of sexual hierarchy rooted in the patriarchal family, and the system of ethnic oppression stemming from racism.

In analyzing the effects on nature of an exploitative society, Reclus showed an awareness of the dangers posed by loss of biodiversity and by ecological disruption that was unusual in his time. In *La Terre*, he presents examples of the extinction of species caused by human "destruction," "slaughter," and "butchery," and concludes that human activity has caused a "rupture in the harmony primitively existing in the flora of our globe."[62] As early as the 1860s, long before wilderness preservation became an organized movement with the establishment of the Wilderness Society in 1936, and indeed even before the establishment of the first national park in the United States in 1872, Reclus was warning of the dangers to ancient forest ecosystems in North America. For example, he laments the loss of "colossal" and "noble" trees like the sequoias of the west coast, which he considers "perhaps an irreparable loss" in view of the "hundreds and thousands of years" that will be necessary for their regeneration.[63] He also discusses the damage produced through the introduction into ecosystems (whether by intention or negligence) of exotic plants and animals without consideration of their effects on the balance of nature. Here again, he focuses on another major ecological problem that has only recently gained widespread attention in "environmental" thought. Reclus quotes the poignant comment of the Maori of New Zealand that "the white man's rat drives away our rat, his fly drives away our fly, his clover kills our ferns, and the white man will end by destroying the Maori."[64]

Despite his remarkable grasp of ecological problems in general, Reclus often shows a great deal less ecological insight in his discussions of demography and population growth in particular. It was his opinion that the human population of one and one-half billion in his time was not only supportable but even "still very minimal, relative to the habitable surface of the earth."[65] He did not seriously consider the impact on the biosphere of such a possibility as several doublings in human population during the next century. At one point, he minimizes the significance of increases in human population by noting that if each person were given a square meter of space, everyone could fit into the area of greater London.[66] Such a fact is, of course, entirely irrelevant from his own standpoint of social geography. We could stand several persons in each square meter, and even put some on the shoulders of others, without learning very much about the interaction between human communities and the earth.

Fortunately, Reclus's discussion of population is often much more nuanced than this, though still tinged with progressivist optimism. He is well aware of the fact that there is no optimal human population that can be calculated by means of arithmetic and plane geometry, or even discovered through more complex natural and social sciences. In this recognition, he was already far ahead of many of our contemporary advocates of simplistic conceptions of "carrying capacity." He notes that if the world consisted of a population of hunters, the earth could perhaps support a population of only five-hundred million, or one-third the actual population as he was writing. He cites various estimates of the possible sustainable human population, and comments favorably on the view of "that circumspect evaluator, Ravenstein," that a population of six billion is a possible limit.[67] However, he expresses skepticism about all such estimates since there are numerous variables that cannot be predicted with any certainty. As an example, he cites changes in methods of production, and, most notably, those in the area of agriculture. In his view, such changes would probably allow a much greater human population to be supported. He believes that when farming attains "the intensive character that science dictates," population will increase at "a completely unforeseen rate," and that "the expanse of good land, which is presently quite limited, cannot fail to grow rapidly, whether through irrigation, drainage, or the mixing of soils."[68] He did *not* stress another set of possibilities that are equally in accord with his basic theoretical orientation: that if vastly increased social and ecological costs of increased technological development lead to a slowing of growth in productivity, if the supply of land dwindles under population pressures, and if ecological degradation causes the quality of the soil to deteriorate, then exactly opposite conclusions concerning population must be drawn.

In reality, Reclus shared with his contemporaries certain pronatalist biases, and saw the decline in birthrates in parts of Europe as a sign of decadence. He moralizes about the fact that in the more affluent areas, natality drops drastically. He cites the examples of the *départements* of l'Eure and Lot-et-Garonne, where the death rate had surpassed the birthrate for most of a century, although these are among the *départements* "whose soils have the greatest fertility."[69] He attributes the failure of the citizens to reproduce at appropriate levels to the egoism of affluence. He sees this failure as an example of the conflict under capitalism between the pursuit of individual self-interest and the general good. He notes that proprietors who fear the division of their land among numerous heirs, and functionaries with modest incomes who want to improve their social status find that having fewer offspring serves their self-interest better.[70] What he fails to note is that where egoism reigns, all social phenomena take on an egoistic coloring, and that their character in such a context says little about these phenomena "in themselves."

Despite his pronatalist tendencies, Reclus did not share the widespread view that increase in population was an unmixed blessing to society. He says that although "growth in numbers has been, without doubt, an element contributing to civilization, it has not been the principal one, and in certain cases, it can be an obstacle to the development of true progress in personal and collective well-being, as well as to mutual good will."[71] Today he would probably see it as an unmixed curse, as ecological devastation accelerates, as the accompanying social crisis intensifies, and as a rapidly increasing human population now approaches the limit of six billion that could seem plausible even in his optimistic age. Moreover, the conditions of production have changed in a sense opposite to the one he hoped for: their development shows little promise of abundance for a rapidly expanding human population, while it threatens to destroy the biotic preconditions for supporting existing human and many other populations at any "optimal level," if indeed at any level at all.[72]

An area in which Reclus was far in advance of his time, and in which he anticipated current debate in environmental philosophy and environmental ethics is in his concern with ethical and ecological issues regarding our treatment of other species. Reclus was unique in being not only a pioneer in ecological philosophy, but also an early advocate of the humane treatment of animals and of ethical vegetarianism. Even today, after several decades of discussion of "animal rights" and "ecological thinking," there are few theorists who have attempted to think through carefully the interrelationship between the two concerns. Yet, a century ago Reclus offered some highly suggestive ideas about how a comprehensive holistic outlook might encompass a serious consideration of our moral responsibilities toward other species.

Reclus observes that all social authorities, in addition to public opinion in general, "work together to harden the character of the child" in relation to animals used for food.[73] This conditioning, he says, destroys our sense of kinship with a being that "loves as we do, feels as we do, and, under our influence, progresses or retrogresses as we do."[74] Like utilitarian defenders of animal welfare since Bentham, he objects to the suffering inflicted on animals raised for food. But adopting a much wider perspective, he also censures the injury caused to the species by the process of domestication. The flourishing and development of species that is possible in the wild is reversed as the animal is increasingly adapted to its single role as a source of food. It has already been noted that Reclus links the ethical and the aesthetic in his analysis of this subject, observing that the abuse of animals that is morally repugnant is also repellent to our sensibilities. He also relates this issue to the question of value. "It is just one of the sorriest results of our flesh-eating habits that the animals sacrificed to man's appetite have been systematically and methodically made hideous, shapeless, and debased in intelligence and moral worth."[75] This reduction of "moral worth" suggests two aspects of the moral problem: first, that humans fail to recognize the intrinsic value or worth of the animal's life and experience; and second, that the "debasing" treatment to which it is subjected reduces the possibilities for the animal's attainment of its own good, or value-experiences. Were Reclus to observe the factory-farming practices of our day, he would no doubt reaffirm this point even more strongly. The importance of ethical vegetarianism, in his view, is that it expresses "the recognition of the bond of affection and goodwill that links man to the so-called lower animals, and the extension to these our brothers of the sentiment which has already put a stop to cannibalism among men."[76]

This reference to "bonds" and "links" indicates how this issue is related to Reclus's general holistic position. In this theoretical context, the issue of treatment of animals goes far beyond the "moral extensionism" of many later theorists who merely adapt conventional, nonecological ethical concepts and apply them to nonhumans. Reclus instead undertakes a fundamental rethinking of the ethical. He believes that our attitude toward other species is not only a question of moral treatment of other *individual* beings, but also a good measure of our awareness of our connectedness to the *whole of nature*. Moreover, an understanding of our relationship to other animals is important in the process of human self-realization, as the domain of reason and that of feeling expand concomitantly. Our growing knowledge of animals and their behavior "will help us to penetrate deeper into the science of life," and "will enlarge both our knowledge of the world and our love."[77] We thus grow morally as the scope of our knowledge grows and as our attachment to the larger system of life is strengthened.

Here as elsewhere in his thought (indeed, going back to his earliest work), the centrality of the concept of love to Reclus's worldview is evident. His view of human moral development is noteworthy in relation to recent discussions of the distinction between an ethics of abstract moral principles and an ethics of care.[78] Reclus is unusual among nineteenth-century radical social thinkers in that he focuses so strongly on the importance of the development of moral feeling, compassion, and the practice of love and solidarity in everyday life. In his time, much of the radical opposition to the dominant order was fueled by a sense of injustice and outrage at the oppression and inequities in society. While this opposition certainly had an authentic ethical dimension, it also succumbed to the reactive mentality and spirit of *ressentiment* that Nietzsche so perceptively diagnosed in socialism, communism, and anarchism. Reclus's outlook achieves a remarkable synthesis between, on the one hand, an interest in justice and the expansion of knowledge and rationality, and on the other hand, a concern for social solidarity and the development of care and compassion. In this synthesis, he anticipates contemporary ethical theorists who seek to restore the balance between these two sets of concerns.

Reclus's conception of love and solidarity is also instructive in relation to issues in contemporary environmental philosophy. While various recent theorists have offered "identification" with nature as an antidote to "anthropocentric" attitudes and practices, such proposals have sometimes remained on a rather abstract idealist level at which identification has the character of an act of will, if not indeed that of a leap of faith. From Reclus's perspective, it is our growing knowledge of (in the sense of both *savoir*, understanding, and *connaître*, being acquainted with) the earth and its human and nonhuman communities that offers an expanded scope for identification and solidarity. As we come to know each realm more adequately, we achieve greater identification with our own species, identification with all the inhabitants of the planet, and finally, as "the conscience of the earth," identification with the living, evolving planet itself.

In this insight, as in so many other aspects of his thought, Reclus anticipated some of the most profound dimensions of contemporary ecological thinking. It is quite striking that a century ago he was exploring in considerable detail so many themes relevant to current fields of interest such as social ecology, ecological holism, animal rights theory, bioregionalism, the ethics of care, and earth-centered narrative. Reclus's social geography, therefore, deserves much greater recognition and continuing study as an important chapter in the history of ecological thought.

Notes

1. None of Reclus's most significant works in social theory have been available in English, and the first collection in English of important selections from his extensive theoretical writings is only now being published. This work also includes the first comprehensive analysis in English of Reclus's social and political thought. For a much more detailed discussion of the issues raised in the present discussion, see John Clark and Camille Martin, eds. and trans., *Liberty, Equality, Geography: The Social Thought of Elisée Reclus* (Littleton, Colo.: AIGIS Pubns. 1996). I would like to thank Camille Martin for her invaluable comments on this chapter.

2. For an introduction to social ecology, see John Clark, ed., "Part Four: Social Ecology," in *Environmental Philosophy: From Animal Rights to Radical Ecology*, ed. Michael Zimmerman et al. (Englewood Cliffs, N.J.: Prentice Hall, 1993). The most extensive presentation of Bookchin's version of social ecology, which is compared with Reclus's social geography at several points in this article, is *The Ecology of Freedom: The Emergence and Dissolution of Hierarchy* (Palo Alto, Calif.: Cheshire Books, 1982). For a spectrum of views associated with social ecology, see John Clark, ed., *Renewing the Earth: The Promise of Social Ecology* (London: Green Print, 1990).

3. Elisée Reclus, *La Terre: description des phénomènes de la vie du globe* (Paris: 1868–69). The first volume was translated as *The Earth: A Descriptive History of the Phenomena of the Life of the Globe* (New York: Harper and Brothers, 1871), and the second as *The Ocean, Atmosphere, and Life* (New York: Harper and Brothers, 1873).

4. (Paris: Hachette, 1876–94), 19 vols. The work was translated as *The Earth and Its Inhabitants: The Universal Geography* (London: H. Virtue and Co., Ltd., 1882–95).

5. Gary S. Dunbar, *Elisée Reclus: Historian of Nature* (Hamden, Conn.: Archon Books, 1978), 95.

6. Elisée Reclus, *L'Homme et la Terre* (Paris: Librairie Universelle, 1905–8), 6 vols.

7. While the emphasis in the present discussion is on the relevance of Reclus's social geography to ecological thought and social theory, the considerable importance of his contribution in other areas, including physical geography and geology, should not be overlooked. Among Reclus's achievements was his early advocacy of the theory of continental drift and his defense of the view that this phenomenon is compatible with uniformitarian explanation. As early as 1872, in *The Earth*, he proposed that the planet is many times older than most contemporary theory indicated, and that the continents formed a single land mass as recently as the Jurassic period. In 1979, an intriguing discussion of Reclus's geological significance appeared in the journal *Geology*. In his article, "Elisée Reclus—Neglected Geologic Pioneer and First (?) Continental Drift Advocate" (*Geology* 7 [April 1979]: 189–92), James O. Berkland concludes that Reclus "was a peer of the geologic greats of the nineteenth century such as Darwin and Lyell" and that while his name "has faded to near obscurity," he "should be recognized in the history of plate tectonic theory as one of its foremost pioneers and perhaps, as its founder" (192). In a comment on this article (*Geology* 7 [September 1979]: 418) Myrl E. Beck, Jr., suggests that Reclus's lapse into "obscurity" may have had more to do with his anarchist philosophy than with the merits of his scientific theories. In his "Reply," Berkland agrees, and laments "the slow literary descent of Reclus to the status of a quasi-nonperson" [sic] as a case of "book-burning through neglect." In his concluding statement, Berkland surprisingly admits that "had [he] possessed full knowledge of just how 'revolutionary' Reclus really was, it is probable that [he] would not have invested the time and effort to give him well-deserved credit for his geologic accomplishments" (*Geology* [September 1979]: 418). I am very grateful to geologist Anatol Dolgoff for drawing my attention to this exchange.

8. Béatrice Giblin, "Reclus: un écologiste avant l'heure?" in *Hérodote* 22 (1981): 110. Giblin edited and wrote the introduction for a book of selections entitled *L'Homme et la Terre—morceaux choisis* (Paris: Maspero, 1982). The entire issue of *Hérodote* containing her article is devoted to studies of Reclus's work, with a strong emphasis on the ecological implications of his social geography. Contemporary ecological thought (with the exception of some varieties of ecoanarchism) has devoted little attention to the connection between geography and ecology. It is noteworthy that in a forthcoming work, Thomas Berry, one of the best-known contemporary ecological thinkers, devotes a chapter to ecological geography, and states that "geography is one of the basic integrating disciplines for those who would enter into ecological studies, with their emphasis on the single community that humans form with the Earth and all its component members" (Thomas Berry, *The Meadow Across the Creek: Ecological Essays* [forthcoming]).

9. Thomas Berry and Brian Swimme, *The Universe Story* (New York: HarperCollins, 1992).

10. "L'Homme est la nature prenant conscience d'elle-même." Elisée Reclus, *L'Homme et la Terre*, 1:1. (All quotations for which the original French edition is cited are my translations, in collaboration with Camille Martin.) The parallel between Reclus's concept and Hegel's idea of human history as a process of Spirit's coming to consciousness of itself is obvious. Indeed, Reclus makes an important contribution to the project of developing a naturalistic, evolutionary reinterpretation of Hegel's conception of "Spirit knowing and enjoying itself as Spirit." It is also instructive to compare Reclus's holistic evolutionary concept to Marx's much less dynamic and holistic conception of nature as "man's inorganic body." While the two thinkers were contemporaries (Reclus being only twelve years younger than Marx), Reclus was much more successful in transcending the spirit of the age by applying a dialectical analysis to the relationship between humanity and nature. For a discussion of Marx's philosophy of nature and his failure to to develop the dialectical naturalism implicit in his thought, see my essay "Marx's Inorganic Body," in *Environmental Ethics* 11 (1989): 243–58, reprinted in Michael Zimmerman et al., *Environmental Philosophy: From Animal Rights to Radical Ecology*, 390–405. It should also be noted that in this area, Reclus far surpassed the contemporary anarchist thinkers, who often shared the limitations of Marx, while lacking the latter's subtlety and complexity.

11. Elisée Reclus, *The Ocean, Atmosphere, and Life* (New York: Harper and Brothers, 1873).

12. Yves Lacoste, "Editorial" in *Hérodote* 22 (1981): 4–5.

13. Yves Lacoste, "Géographicité et géopolitique: Elisée Reclus" in *Hérodote* 22 (1981): 14. While American geography once accorded Reclus a significant level of recognition, it has engaged in a similar process of "forgetting." For example, one finds in *The Geographical Review* (founded in 1916) three references to Reclus in the 1920s, three in the 1930s, two in the 1940s, and then a long silence. There was a modest resurgence of interest in Reclus among American geographers during the 1970s. This is evidenced by articles dealing with his work in the radical geography journal *Antipode*, and the publication of geographer Gary Dunbar's biography *Elisée Reclus: Historian of Nature* (Hamden, Conn.: Archon Books, 1978).

14. Reclus, *L'Homme et la Terre*, 6: 504.

15. Reclus, *L'Homme et la Terre*, 1: 14.

16. Reclus, *L'Homme et la Terre*, 6: 504.

17. Reclus, *L'Homme et la Terre*, 6: 527.

18. Reclus, *L'Homme et la Terre*, 6: 527.

19. In Reclus's time, just as today, there were views that overemphasized unity and the whole and others that overemphasized diversity and the individual phenomenon. In the past century, much of the organicist tradition stemming from Hegel tended toward extreme holism and social authoritarianism, while the individualist tradition arising out of classical liberalism produced social atomism and anomic individualism. An authentically dialectical position,

which interprets the whole as a dynamic, developing unity-in-diversity, avoids both of these dangers without resorting to *ad hoc* solutions to internal contradictions. For a discussion of the ecosystem model of Clements and Odum, with its implications of order, harmony, and homeostasis, and later challenges to that model, see Donald Worster, "The Ecology of Order and Chaos," in *Environmental History Review* 14 (1990): 1–18. For more extensive treatment of the history of ecosystems theory, see Worster's *Nature's Economy: A History of Ecological Ideas* (New York and Cambridge: Cambridge University Press, 1994), esp. chap. 16.

20. Reclus, *The Ocean*, 434.
21. Reclus, *L'Homme et la Terre*, 6: 254.
22. Reclus, *L'Homme et la Terre*, 6: 255.
23. Reclus, *L'Homme et la Terre*, 6: 255. Reclus's holism may be compared to a similar strain in the thought of his friend and colleague Kropotkin, who contends that geography should "represent [nature] as a harmonious whole, all parts of which are . . . held together by their mutual relations." See "What Geography Ought to Be," quoted in Myrna Breitbart, "Peter Kropotkin, Anarchist Geographer," in *Geography, Ideology and Social Concern*, ed. David Stoddart (Oxford: Blackwell, 1981), 145. There are also striking similarities between Reclus's views and the Gaia hypothesis. Reclus's description of the earth "regulating" itself through forests may be compared to James Lovelock's definition of Gaia as "a complex entity involving the earth's biosphere, atmosphere, oceans, and soils; the totality constituting a feedback or cybernetic system which seeks an optimal physical and chemical environment for life on this planet" (*Gaia: A New Look at Life on Earth* [Oxford: Oxford University Press, 1979], 10).

24. Thérèse Dejongh, "The Brothers Reclus at the New University," in *Elisée and Elie Reclus: In Memoriam*, ed. Joseph Ishill (Berkeley Heights, N.J.: The Oriole Press, 1927), 237.

25. While Bookchin has used the terms "first nature" and "second nature" frequently in recent years, he never presents a detailed philosophical analysis of the relationship between the two realms. In his essay "Thinking Ecologically," he states that by "second nature" he means "humanity's development of a uniquely human culture, a wide variety of institutionalized human communities, an effective human technics, a richly symbolic language, and a carefully managed source of nutriment" (Murray Bookchin, *The Philosophy of Social Ecology* [Montréal: Black Rose Books, 1990], 162). He describes "first nature" as the larger natural world from which second nature is "derived." "The real question," he says, "is how second nature is *derived* from first nature" (163). Unfortunately he does not go very far in answering this key question. He also posits a third natural realm, called "free nature," which he does not describe as an existent sphere, but rather as a possibility in a future ecological society. He says that it would constitute "a nature that could reach the level of conceptual thought" (182). This is, however, a confused formulation, since

nature has already reached "the level of conceptual thought" in what he calls "second nature."

26. Reclus, *L'Homme et la Terre*, 1: 42.
27. Reclus, *L'Homme et la Terre*, 1: 42.
28. Reclus, *L'Homme et la Terre*, 1: 117.
29. Reclus, *L'Homme et la Terre*, 1: 117.
30. Reclus, *L'Homme et la Terre*, 1: 119.
31. Reclus, *The Ocean*, 435. Of course, a bioregional perspective is not merely descriptive, although it may begin with an analysis of how our natural regions shape our selves and communities. The point of bioregionalism is to generate a creative dialectic between culture and place. The sense of place is a poetic response to nature on the part of the human imagination. The best sources on bioregionalism are the works of Gary Snyder and the publications of the Planet Drum Foundation and other bioregional organizations. See Snyder's "The Place, the Region, and the Commons" in *The Practice of the Wild* (San Francisco: North Point Press, 1990), 25-47, and *A Green City Program* (San Francisco: Planet Drum, 1989). For the place of regionalism in the ecology of the imagination, see Max Cafard, "The Surre(gion)alist Manifesto," in *Exquisite Corpse* 8 (1990): 1, 22-23.
32. One of the many similarities between the social geography of Reclus and that of Kropotkin lies in the strongly bioregional flavor often found in the works of both. Myrna Breitbart in "Peter Kropotkin, Anarchist Geographer" points out that he "believed that it was necessary to reestablish a sense of community and love of place. Rootedness in a particular environment would foster greater human interaction and a more intimate relationship with one's surroundings" (140).
33. Breitbart, "Peter Kropotkin, Anarchist Geographer," 140.
34. See Baron de Montesquieu, *The Spirit of the Laws* (New York and London: Haffner Publishing Co., 1949), chaps. 14-17. Neither should it be confused with the work of a historian such as Le Roy Ladurie, whose impressive study *Times of Feast, Times of Famine: A History of Climate Since the Year 1000* (New York: Farrar, Straus and Giroux, 1971) deals—as the title indicates—with the effect of the vicissitudes of climate on a variety of social conditions. Reclus should rather be compared with thinkers who investigate the effect of the constants of climate on the character of cultures and peoples. More in his tradition is critic and photographer D. E. Bookhardt, who writes that the great Louisiana surrealist photographer Clarence Laughlin "attempted to confront the *genius loci* head on. Relics of the cultural landscape, subjected to the ferocity of the subtropical elements over time, served as foils for his visual reveries—a kind of Old South vision of Atlantis, infested with ghosts and creatures of indeterminate mythology, all illumined by a spectral, tropical radiance" ("The Jungle is Near: Culture and Nature in a Subtropical Clime," in *Mesechabe* 2 [1988-89]: 4).
35. Ellsworth Huntington argues that there is "a close adjustment between

life and its inorganic environment," and that factors such as "soil, climate, relief" and "position in respect to bodies of water" all "combine to form a harmonious whole" in affecting human society" (*The Human Habitat* [New York: D. Van Nostrand, 1927], 16–17). It turns out that this "harmonious whole" dictates racial hierarchy, since "racial differences" in areas such as "inherent mental capacity" are caused by the various natural factors, especially climate ("Climate and the Evolution of Civilization," in *The Evolution of the Earth and Its Inhabitants* [New Haven: Yale University Press, 1918], 148). Elsewhere he seeks to defend his racialist conclusions by arguing—or more accurately, speculating—that climate has had an enormous influence on inheritance through its effects on "migration, racial mixture, and natural selection," and perhaps even "mutations" (*Civilization and Climate* [New Haven: Yale University Press, 1915], 3).

36. Reclus, *L'Homme et la Terre*, 2: 91.
37. Reclus, *L'Homme et la Terre*, 2: 91.
38. Reclus, *L'Homme et la Terre*, 2: 91.
39. Reclus, "Du Sentiment de la nature dans les sociétés modernes," in *Revue des deux mondes* 63 (1866): 379.
40. Thomas Berry, "The Viable Human," in *Environmental Philosophy: From Animal Rights to Radical Ecology*, ed. M. Zimmerman, et al., 174.
41. Elisée Reclus, "Du sentiment de la nature," 379–80.
42. Elisée Reclus, "Du sentiment de la nature," 379–80.
43. Elisée Reclus, "On Vegetarianism," in *The Humane Review* (January 1901): 318.
44. Reclus, *The Ocean*, 526.
45. Reclus, *The Ocean*, 526.
46. Reclus, *The Ocean*, 526.
47. Reclus, "On Vegetarianism," 322.
48. Reclus, "On Vegetarianism," 323.
49. Reclus, *The Ocean*, 526.
50. Reclus, *The Ocean*, 527.
51. Reclus, *The Ocean*, 527.
52. Reclus, "Du sentiment de la nature," 379.
53. Reclus, "Du sentiment de la nature," 379.
54. "Free nature" is used in this case in Arne Naess's sense of areas in which spontaneous ecological processes can take place without major human disruption. George Sessions claims that social ecologists "have yet to demonstrate an appreciation of, and commitment to, the crucial ecological importance of wilderness and biodiversity protection" ("Wilderness: Back to Basics," an interview by JoAnn McAllister with George Sessions, in *The Trumpeter* 11 [Spring 1994]: 66). Yet, a dialectical, holistic position that sees humanity as "the self-consciousness of the earth, " interprets history as the movement toward a "free nature" (in a sense that synthesizes Naess's and Bookchin's concepts), and conceives of the earth as a unity-in-diversity, is

eminently capable of dealing theoretically with these important issues. Steve Chase has presented a very circumspect analysis of the neglect of wilderness issues by Bookchin and many other social ecologists, and the need for attention to these issues from a social ecological perspective. See "Whither the Radical Ecology Movement?" in *Defending the Earth,* ed. Steve Chase (Boston: South End Press, 1991), 7–24.

55. Reclus, *The Ocean,* 529.
56. Reclus, *The Ocean,* 529.
57. For a perceptive discussion of "otherness" and the distinction between "splitting" and "differentiation" in relation to the other (including nature as other), see Joel Kovel, *History and Spirit: An Inquiry into the Philosophy of Liberation* (Boston: Beacon Press, 1991), 45–58.
58. This raises an important issue not only for Reclus, but for social ecology. While humanity can and ought to make a unique contribution to the emergence of greater freedom and creativity in nature, this contribution cannot be limited to humanity's attainment of its own nondominating self-realization and to creative interaction with the natural milieu in a way that respects the integrity of nature, as important as these goals may be. At this point in the history of the earth, another essential ecological question is the way in which human beings can reorganize society so that its impact on large areas of the earth can be reduced and finally minimized. A stronger conception of "nondomination" is needed: one that recognizes the need for the earth to have a sphere of ecological freedom and evolutionary creativity guided neither by human self-interest nor by human rationality.
59. Elisée Reclus, "The Progress of Mankind," in *The Contemporary Review* 70 (July–December 1896): 782.
60. Reclus, "The Progress of Mankind," 782
61. *Homme et la Terre,* 6: 538. It is not clear precisely to what extent Reclus believed that the elimination of social domination would result in an end to human antagonism toward the natural world. He never states in a simplistic, undialectical way that the former is a necessary and sufficient condition for the latter. As a general principle he thought that the establishment of a society based on cooperation, love, and aesthetic appreciation would result in nonexploitative institutions and patterns of behavior in relation to humanity, to other species, and to the larger natural world.
62. Reclus, *The Ocean,* 517–18.
63. Reclus, *The Ocean,* 518.
64. Reclus, *The Ocean,* 519; quoted by Reclus, who cites Hasst, von Hochstetter, and Peschel in *Ausland* (19 February, 1867).
65. Reclus, *L'Homme et la Terre,* 5: 300.
66. Reclus, *L'Homme et la Terre,* 5: 300.
67. Reclus, *L'Homme et la Terre,* 5: 332.
68. Reclus, *L'Homme et la Terre,* 5: 332.
69. Reclus, *L'Homme et la Terre,* 5: 415.

70. Reclus, *L'Homme et la Terre*, 5: 416.
71. Reclus, *L'Homme et la Terre*, 5: 418.
72. This is not to deny the obvious fact that scarcity is socially generated, and that thus far the burden of famine and malnutrition has fallen most heavily on those who suffer from economic and political powerlessness, not on those who "overuse resources" most flagrantly. See Frances Moore Lappé and Joseph Collins, *World Hunger: Twelve Myths* (New York: Grove Weidenfeld, 1986). The point is, however, that from a global ecological perspective, continued population growth will necessarily aggravate ecological crisis, whatever other social variables may exist.
73. Reclus, "On Vegetarianism," 318.
74. Reclus, "On Vegetarianism," 318.
75. Reclus, "On Vegetarianism," 318.
76. Reclus, "On Vegetarianism," 322. Reclus's arguments constitute an eloquent defense of the humane treatment of animals, but they are far from conclusive as a proof of the moral necessity of strict vegetarianism. He presents an excellent case for the immorality of systems of food production that inflict continual suffering on animals and callously ignore the moral relevance of the attainment of goods or of the self-realization of these beings. His critique would therefore apply to much of today's meat industry, with its factory farming and mechanized mass production. In addition, his arguments concerning the evils of domestication present a strong case against raising certain animals even in nonfactory conditions. Nevertheless, he does not demonstrate that *all* forms of animal husbandry and hunting are inhumane. It is noteworthy that Reclus never subjects traditional hunting societies to the scathing criticism he directs toward the modern meat industry. Unfortunately, he fails to explore the possibility of morally relevant differences between the two systems.
77. Elisée Reclus, "The Great Kinship," trans. Edward Carpenter, in *Elisée and Elie Reclus: In Memoriam*, ed. Joseph Ishell, 54.
78. See Carol Gilligan, "Moral Orientation and Moral Development" in *Women and Moral Theory,* ed. Kay Kittay and Diana Meyers (Totowa, N.J.: Rowman & Littlefield, 1987), 19–33. According to Gilligan, "since everyone is vulnerable to both oppression and abandonment, two moral visions—one of justice and one of care—recur in human experience" (20). This essay develops further the ethical implications of her groundbreaking work, *In a Different Voice: Psychological Theory and Women's Development* (Cambridge, Mass.: Harvard University Press, 1982).

The Maintenance of Natural Capital: Motivations and Methods

Clive L. Spash and Anthony M. H. Clayton

Concern for environmental degradation has lead to the suggestion that natural capital maintenance is a necessary but insufficient condition for a sustainable society. Natural capital tends to be discussed in an all-inclusive sense without explicitly recognizing the attributes that make it so important. In the following pages the reasoning behind concerns for natural capital maintenance are evaluated and the methods that claim to achieve that goal are discussed. The spectrum of concerns ranges from the anthropocentric utilitarian perspective of neoclassical economics, where natural capital is required to provide income flows, to the role of preventing the loss of critical functions and maintaining stability in ecosystems. This paper moves through the range of perspectives on natural capital to offer a view of what is being discussed by different schools of thought.

Those who regard ecosystems as fundamentally robust, and environmental problems as minor perturbations, view the warnings of neo-Malthusian, environmental pessimists (which drive sustainability concerns) as pure fantasy because they lack adequate scientific evidence. However, the application of risk, uncertainty, and ignorance to the decision-making process results in the realization that the standard approach to the economic modeling of uncertain future events suffers from a narrow reductionist view of knowledge. Rather than continuing the search for an optimal path forward, to which society can be irreversibly committed, the recognition of our ignorance emphasizes the need to preserve flexibility by opening out the selection of options.

The tendency has been to restrict the decision process to a few ways forward as suggested by a simple linear extrapolation of the past.

An economic view of sustainability in terms of value is confronting an ecological view of sustainability in terms of physical characteristics. The restricted economic outlook has been shown to fail at assuring stability, which is regarded as ecological sustainability, while being able to maintain constant consumption, taken as economic sustainability.[1] Thus, the sovereignty of the consumer can be seen as a potential threat to the general system, which must therefore be restricted by allowing the requirements of the system to override those of the consumer. The combination of ecological and economic approaches raises issues of a more fundamental philosophical nature. These philosophical issues seem to motivate much environmental concern, and are shown below to be directly relevant to the position taken by some prominent environmental economists.[2]

Both the ecological and economic approaches require that the factors to be protected be identified, and that natural capital be shown as a cause of that protection, suggesting the issue were one of scientific determinism. However, the definitions of natural capital are then a function of the basic philosophical assumptions of value formation. In particular, adopting a nonanthropocentric perspective recognizes that elements of natural capital are of value in themselves and cannot be captured by economic valuation. This has direct relevance to the strategies put forward for natural capital maintenance and their ability to address the concerns that seem to lie behind the need to have a concept of natural capital at all. Once these arguments have been developed, four approaches to natural capital maintenance are discussed: economic valuation, physical compensation, scientific thresholds, and a systems approach.

Why Natural Capital?

M. Redclift has argued that the mode of thinking summarized in modern economics (i.e., increasing material throughput maximizes welfare) is largely responsible for the unsustainable development of both North and South.[3] In his view, the pursuit of growth, and neglect of its ecological consequences, has its roots in the classical paradigm that informed both market and socialist economies. Others point to a number of human-induced environmental problems to show that society needs to plan explicitly for sustainability.[4] In contrast, the neoclassical economic viewpoint claims that, even without technological progress, nonrenewable resource depletion still allows sustainable development because economic output can be maintained or even increased via substitution.[5] Yet, within this framework, rapid resource depletion increases the dependence of the current system upon technological progress to prevent collapse when substitutes are limited. In addition, economic models have erroneously reflected the assumed nonscarcity of environmental source and sink functions.[6] This assumption is directly acknowledged as inadequate where the common value of such functions are excluded from the selfish calculations of economic persons.

The standard economic model of production includes land, labor and capital. Following Marshall, land is taken to include all "free gifts" of nature and is assumed to be fixed. That is, land is the input to production that humans cannot increase (unlike man-made capital), but only utilize more efficiently. If the concept of natural capital is defined within the neoclassical utilitarian framework, the justifications for its consideration as separate from other capital lie in the degree of substitution between types of capital. Natural capital would therefore seem to be neoclassical land renamed. This could be worthwhile because land was more relevant as an input to an agriculturally based economy while the new name implies use of a wide range of natural resources. However, if natural capital is merely semantics, little in the way of new insights to sustainability can be expected. In fact, the application of such a restricted definition raises several problems that show this is far from being the extent of concern.

Definitions of natural capital in the literature tend to be vague at best. Typically reference is made to a stock of resource and environmental assets.[7] Natural capital is taken to include those features of nature (such as minerals, biological stocks, and pollution absorption capacity) that are directly or indirectly utilized, or are potentially utilizable, in human social and economic systems. This recognition of indirect values comes from the ecological concern for those features (such as soil and atmospheric structure, plant and animal biomass) that form the basis of ecosystems.[8] The concept of natural capital raises a concern for the indirectly utilized features of nature in a similar way to Boulding's emphasis on the value of stocks, as opposed to flows, in the creation and maintenance of human systems and their improvement over time.[9]

While conceptualizing the stock of natural capital is an issue for all concerned with its maintenance, this is particularly problematic for the economic approach. In fact, the need to define the economic boundaries of this stock has been obscured by focusing on the maintenance of income flows from capital rather than being concerned for the measurement and meaning of the total stock. For example, Serafy has argued that efforts should be concentrated upon measuring income flows, while leaving aside the valuation of the total environment, in order to obtain a "sustainable" future income.[10] However, income flows cannot be separated from the capital stock (and embodied technology) from which they derive. Under this definition, the value of natural capital lies in the amortized value of future benefits that can be derived from the use of the asset.[11] The aim of natural capital maintenance is then to maximize the capital yield in net present value terms. This requires avoiding diminution of natural resource stocks and deterioration and degradation of the environment only if technological limits or lack of substitutes make this a least cost strategy. An important differentiation is however necessary in terms of income flows from capital stock reductions, running down natural wealth, as opposed to income from capital yield (only the latter being consistent with long run sustainability). The former is undesirable because the generation of income by the erosion of capital stock tends to reduce future productivity, although this may be regarded as unproblematic in models where the future is discounted.

In terms of sustainability criterion, the concern is to achieve a nondeclining income flow from capital, which maintains or increases utility. If natural capital is reduced, man-made capital will need to compensate for the yield lost. Thus, the Hartwick rule suggests achieving intertemporal efficiency in resource allocation by investing depletable-resource rents in man-made capital, and so maintaining a constant consumption stream.[12] The constant consumption stream is justified by an appeal to a "Rawlsian" approach to intergenerational justice; an approach to intergenerational fairness suggested for adoption in economics by Solow.[13] However, the simple Hartwick rule depends upon man-made capital: (1) failing to depreciate; (2) being a substitute for, rather than a complement to, natural capital; and (3) being unrelated to rather than produced from natural capital.[14] Ignoring these problems, maintaining the value of the stock of capital is a necessary condition for what Common and Perrings call Solow-sustainability and therefore measures of the value of natural capital are still required.[15]

The inseparability of the definition of sustainable income flow from the natural capital stock is similar in circularity to the attempts at defining an index for man-made capital. The debate, between neoclassical and post-Keynesian economists, concerning this problem is known as the Cambridge Controversy and seems relevant to the desired measurement of natural capital. The difficulty is how to define a measure of aggregate capital when the valuation of capital presupposes a particular interest rate but the interest rate is dependent upon the marginal product of capital, which varies with the quantity of capital; a circular argument arises.[16] In fact, once the term "capital" is taken to apply to the environment, all the issues related to man-made capital seem to be relevant to natural capital.[17] This runs the danger of implying that the environment is actually a subset of man-made capital. If anything, the reverse must be true because natural resource inputs are necessary for the production of man-made capital. Thus, we need to identify those aspects of natural capital that lie behind its adoption by the environmentally concerned community, and that also do differentiate it from both man-made capital and the neoclassical conception of land.

The concern for stability is perhaps the defining physical characteristic that makes natural capital different from man-made capital. The difference turns on the degree to which natural capital is critical to the human species. Something that is critical for one species may of course be noncritical for another. An event that eliminated the human species but left anaerobic life forms otherwise unaffected could be judged as a noncritical loss of natural capital from the point of view of anaerobic life forms. As rational, economic, human beings our primary concern for criticality refers to the potential implication for ourselves of the loss of some particular form of natural capital. Adopting this anthropocentric viewpoint at a societal level, the loss of natural capital is only crucial in so far as damage occurs either to the economic system or the survival of the human species.

Typically critical limits can be reached via human actions in the following ways:

1. Persistently harvesting a renewable resource at a rate that exceeds the maximum sustainable yield; so reducing the stock of the resource until it becomes extinct. In this way a renewable resource might be regarded as moving from the noncritical into the critical category. However, this would only be a threat to critical limits to the extent that the species were either economically important or an essential part of ecosystem stability. As standard texts in resource economics show, extinction of species can under certain circumstances be perfectly rational and even optimal.

2. Eroding the assimilative capacity of the environment to the stage

where serious threats to economic and human systems occur. In this context, a reforestation program with the aim of carbon fixing could be viewed as an improvement of global natural capital—increasing stability and reducing criticality.

3. Depleting a nonrenewable resource that has no substitutes and upon which society depends. This type of concern drove the debate over British dependence upon coal at the turn of the century.[18] The modern concern is where a nonrenewable resource is critical to human systems in an ecological sense and as a result there is no option but to maintain the current stock, for example, stratospheric ozone.

This last point raises a distinction between substituting for exhausted elements of natural capital in economic processes, and substituting for these same elements of natural capital in ecological processes. Ignorance of ecological thresholds means human society could approach a threshold of system tolerance while the element of critical natural capital being depleted was (in economic terms) still relatively abundant. At such a point there might be little direct economic pressure to seek and develop substitutes, although there would be clear ecological reasons for doing so.[19] Yet the uncertainty over potential outcomes is used by the anti-environmentalist lobby to argue that ecosystems are robust until science proves otherwise.[20] This criticism is worth addressing directly because natural capital starts to become more relevant once both uncertainty and ignorance are considered.

Uncertainty and Ignorance

A case against the view that has been behind much environmental concern, variously referred to as "limits to growth," "ecodoom," or "neo-Malthusian," is based on an assessment of the probability that the view is incorrect. A dominant approach to an uncertain world is to try to reduce potential future states to probabilistic events. This requires the estimation of the risk associated with every possible predicted outcome. Practitioners and advocates of this methodology are required to make several questionable assumptions, which include:

1. A cause and effect relationship can be established to determine the outcomes to be included in the set of possible future states. This is often impossible when dealing with sustainability because some of the connections between the dimensions of the sustainability problem are poorly understood.

2. Probabilities are assumed to be associated with all future states

of the world, although, the action leading to an event may be recognized as a possible state without a probability being attached to the outcome. Thus, an event can be expressed as uncertain yet have no associated probability of occurrence. The probability itself may be unknown or nonexistent. (Such a division of risk and uncertainty can be found in Keynes.)[21]

3. The type of partial knowledge being analyzed is assumed to be the risk associated with the occurrence of outcomes. However, all the models of the behavior of complex systems, such as environmental and economic systems or their interactions, are imprecise and limited in their scope. These limitations arise for a number of reasons: ignorance about a particular system, ignorance about the behavior of a class of systems, and the indeterminate nature of some complex systems (which can become chaotic at various points). This means the behavior of such systems can only be modeled in probabilistic terms, for limited domains, or for a limited time.

4. The distribution of risk over space and time is assumed unimportant when judging appropriate action. Yet, many decisions involve choosing between options that have different risks for different people at different times. For example, a small risk of a major disaster (e.g., a 0.01 percent chance of a major disaster killing ten thousand people) compared to a large risk of a small disaster (e.g., a 100 percent chance of killing one person) can give the same expected outcome in terms of expected losses (one statistical life). In practice, people are sensibly concerned about the distribution of such risks. These types of decisions involve making judgments as to the priorities of society. The choice of the appropriate course of action, given a full and accurate picture of the risks associated with different outcomes, would appear to be a political and moral issue. Although, Chichilnisky and Heal suggest how communal institutions could internalize risks faced by individuals via mutual insurance contracts.[22] They also, by reducing ignorance to uncertainty over event frequency distributions, claim state contingent securities can cover global risks such as climate change.

There are, however, some areas of ignorance that cannot be easily placed into the framework of knowledge about systems, as points 1 to 3 suggest. For one useful categorization and explanation of types of ignorance, see Faber et al.[23] In general, where altering the potentialities of systems causing changes that are, in principle, unpredictable, the appropriate response is to maintain options. This implies accepting the importance of different views on the same problem and questioning current knowledge. Stirling argues that a rigorous approach to ignorance is feasible and, learning from operations research, would emphasize criteria of flexibility and reversibility.[24] Walters has argued for a head-on approach to issues of uncertainty, that ignorance needs to be recognized as a first step to knowledge, and that resource management must be adaptive in the face of ignorance.[25]

Now reconsider the role of natural capital and how it fits into the probabilistic framework. Natural capital could be regarded as an insurance premium against known but uncertain future states of the world, where the probability of those states occurring is known or knowable. This would be consistent with an expected utility framework, and could justify a safe minimum standard approach. If the definition of natural capital as critical is adopted, the economy could be "safely" allowed to erode "land" to a hard core of what humans hope they have identified correctly as the essential elements. Furthermore, technology may substitute natural capital with man-made capital over time as critical functions fall within human capacity to produce. However, even if this utilitarian model and its consequences for how the world is viewed are accepted, caution is required in the manipulation of natural capital.

First, there are elements, substances, and organisms on the planet that have *not yet been utilized directly by humans*. This can be viewed as uncertainty and ignorance over future use patterns. For example, technology might enable the use of a previously untapped or uneconomic resource, or research into causes of disease (e.g., cancer) might lead to the recognition of higher-value uses of current resources; adaptation in a dynamic world emphasizes the importance of diversity.

Second, many of the features of nature that are directly utilized in economic processes are *dependent on features of nature that are indirectly utilized*. Current biomass depends on an ecological infrastructure that enables flows into human systems but is ignored itself. The sustainable harvest rate of a given species of fish, for example, will depend on the maintenance of the complex web of relationships that constitute the proximate ecology of that species. The sustainability of the harvest rate, then, depends on the way in which that resource is used. Such use patterns can be relatively direct, for example, the way in which the species itself plus prey, predator, and competitor species were caught. However, "use" patterns can also be quite indirect, and therefore less obvious. For example, emissions of chlorofluorocarbons (CFCs) might reduce available fish catches. That is, stratospheric ozone can be depleted by CFCs so allowing higher levels of UV-B radiation to reach the surface of the planet, which would in turn affect the marine biota at the base of the food chain on which the harvested species of fish depends. In this way, uncertainty and ignorance pertain to ecosystems functions in addition to risk.

Once the above arguments are accepted, an optimal level of the insurance premium would be undefinable. Thus, while natural capital can be defined as merely an insurance premium, this definition requires the rejection of wider concepts of uncertainty and of ignorance. A constant natural capital stock is now motivated by the need to accept our ignorance, which is in accord with setting ecological constraints upon the economic system.

Admitting there are dimensions and elements of natural capital that generate utility but are unknown or unknowable is still in accord with the use of utilitarianism. The main basis for defining natural capital so far, in this paper, has been neoclassical utilitarian. That is, the consequences of depleting natural capital were judged in terms of its value, and that value was purely based upon anthropocentric utility or satisfaction. This is true from either economic or ecological perspectives. The ecological approach is fundamentally utilitarian. As described by Foy, the aim is the definition of physical constraints on the economic system to ensure the sustainability of that system.[26] The constraints based on ecological criteria are concerned with the instrumental value of other species and the environmental systems for human life. However, Foy also notes and alludes to, but fails to discuss, a second set of constraints based upon ethical considerations of duties to future generations and other species.

Our discussion has also excluded consideration of other living beings that are elements of natural capital. This raises questions as to the extent to which human-animals can legitimately regard such elements as resources for their use, which in turn depends upon the moral standing of nonhuman beings, the extent to which they can feel pain, and the extent to which this is acknowledged. More generally, natural capital as defined so far is an inadequate expression of the components of the world system, and the justifications for its maintenance are being driven by more than economic and ecological considerations.

Beyond the Utility of Natural Capital

A utilitarian analysis of natural capital is dependent upon the value of that capital in its current state as opposed to in an alternative use. That is, the use of natural capital is determined by its instrumental value. Economics assumes such valuation of alternatives can be carried out via assessing the willingness-to-pay (or willingness-to-accept compensation) of individuals. Thus, at some level the cost of preserving natural capital will exceed the willingness-to-pay for that use and it will be depleted, thereby maximizing utility. Such a process may be perfectly reasonable from a utilitarian perspective, but if society accepts the existence of values in nature that are outside of the human calculus, a conflict will arise. For a society aiming to protect such natural values, willingness-to-pay is a redundant concept and natural capital will take on a different meaning. The difference can be regarded as the realization of an intrinsic value in nature as well as in humans (i.e., some values are not anthropocentric although they are recognized by humans).

A utilitarian philosophy sees only instrumental value in acts but intrinsic value in the consequences of those acts. Human welfare, or happiness, is then seen as the only intrinsically valuable thing: an anthropocentric value system. Under this anthropocentric view, all other things are valuable only insofar as they serve to increase human welfare. The rightness or wrongness of an act is determined by the results that flow from it. Under Bentham's philosophy, pain is bad, pleasure good, and acts can be judged on the net outcome; a good act is one creating more pleasure than pain.

Preservation of natural capital under the utilitarian value system is judged by the results in terms of human welfare. Thus, the reasons for conserving specific sites (e.g., old growth forest in the Pacific Northwest, or Scottish peat bogs) will include the potential for scientific research, maintenance of genetic diversity for medicine and agriculture, recreation, solace, and aesthetic enjoyment. These instrumental values by their influence on human welfare suggest the potential for the economic analysis of preservation benefits. Maintenance of natural capital is then only one possible alternative use of the site and must be weighed against others that may provide greater human welfare. For example, in the United Kingdom, as the need for more roads increases, the more Sites of Special Scientific Interest (areas protected on conservation grounds and nominally out of bounds to development) will be developed, unless the utility value they posses increases.[27]

This raises many issues concerning environmental valuation, cost-benefit analysis, obligations to other species or generations, and most fundamentally the potential for trade-offs. Maintenance of natural capital is but one goal in society and can, under a utilitarian philosophy, be overridden by other human interests. Where the value of a specific type of natural capital, compared to development use, is deemed relatively low, the site will be destroyed by roads, housing estates, or resource extraction. Thus, supporters of the utilitarian philosophy, such as Passmore, cannot, and would not wish to, preclude any area from some eventual development.[28]

This anthropocentric view of the world can be broadened by the inclusion of animals in the utilitarian calculus. Indeed Bentham saw their inclusion as a part of the utilitarian system that he proposed.[29] The "greatest happiness" included avoiding pain and suffering of animals and creating pleasure for them. As long as animals can suffer, avoidance of animal suffering increases utility. Within this structure, a hierarchy of sensitivities has been suggested, attributing the highest sensitivity to humans. In terms of natural capital maintenance, the implications of including the utility of animals would undoubtedly be extensive. However, there is no easy way to estimate the preferences of animals or assess their willingness-to-pay! The idea of including animals starts to move economics into the realm of moral philosophy, and deep ecology.

A step beyond the utilitarian (including nonhuman-animals) argument is the appeal to rights, deontological ethical theories, and intrinsic value in things as well as humans. This is reflected in Aldo Leopold's land ethic, which implies a basic right of natural beings to continue existing in a natural state. As Leopold states:

> The "key-log" which must be moved to release the evolutionary process for an ethic is simply this: quit thinking about decent land-use as solely an economic problem. Examine each question in terms of what is ethically and esthetically right, as well as what is economically expedient. A thing is right when it tends to preserve the integrity, stability, and beauty of the biotic community. It is wrong when it tends otherwise.[30]

The concept of "rights" for flora and fauna can form an absolute constraint on various forms of action regardless of the benefits. Rights operate to provide those individuals or things that hold them with moral standing. That is, status is an end in itself rather than a means to an end. Deontological ethical theories attribute intrinsic value to features of acts themselves. Respectful treatment of natural entities and natural systems would then rule out certain types of exploitative acts on deontological grounds.[31] The use of natural entities and systems as objects and resources of instrumental value could be precluded on grounds of respect and the obligation of noninterference in anything with internal self-direction and self-regulation.

As mentioned earlier, something that is noncritical for human life could easily be critical for nonhumans. There are then two aspects to the argument for natural capital maintenance: (1) the instrumental values recognized by utilitarianism, including ecological maintenance for humans and (2) intrinsic values and ecological maintenance for nonhumans. There are, of course, many issues underlying the recognition of nonhuman animal rights, the foremost of which is the problem of conflicts of interest. One suggestion in this case is that the idea of nonhuman rights can be made more generally acceptable if it is based upon interests, and allowance is made for ranking of rights.[32] The point to be made here is that recognizing the general concepts encapsulated by "rights" is an important motivation for the belief in constant natural capital stock, and nonhuman intrinsic value is implicit in the stance of economists arguing for constant natural capital.

Holland has argued that intrinsic values lie behind the position of authors pushing for constant natural capital from the social scientific approach, and in particular Pearce et al.[33] The latter authors' argument for constant natural capital is supposedly derived from a sustainability criteria based upon justice for future generations. However, constant natural capital is argued by Holland to be an implausible logical outcome of a theory of sustainability based exclusively on the aim of securing justice for future generations; a point that also follows from Common and Perrings.[34] The justification for natural capital is apparently nonhuman intrinsic value. However, Turner and Pearce claim (in a paper not cited by Holland) that by taking care of justice for future generations of human-animals, the concerns of future nonhuman animals will be largely met.[35] There is no need to acknowledge that the latter have intrinsic values, which the authors regard as radical and a waste of economists' time. As Turner and Pearce state: "We have argued that an ethic 'for the use of the environment', which restricts rights to humans and recognises primarily only instrumental value in nature, can in any case offer sufficient safeguards. More progress may be made if the analysts turned their attention to the individualist basis of utilitarianism and conventional economics."[36] More than this, the concept of intrinsic values in non-humans is positively dangerous for three reasons: "(i) it is stultifying of development and therefore has high social costs in terms of development benefits foregone; (ii) it is conducive to social injustice by defying development benefits to the poorest members of the community, now and in the future; (iii) it is redundant in that the modified sustainability approach generates many of the benefits alleged to accrue from the concern for intrinsic values."[37]

This modified sustainability approach actually claims to hold a physical stock of natural capital constant by compensating projects (discussed in more detail in the next section). This constancy of the stock must then imply exclusion from economic development criteria, such as cost-benefit analysis, but inexplicably avoids criticisms (i) and (ii) above. Turner and Pearce also face the problem of finding a justification for a constant physical capital rule purely on grounds of economic efficiency and intergenerational equity (i.e., Holland's arguments still apply). None of the points are explained, so the reader is left wondering: what are these benefits in point (iii), how are these benefits met, and why bother with them if the authors can allege their nonexistence?

In fact, this rejection of nonhuman intrinsic values is strangely qualified by the apparent need to make the concept fit into the economic utilitarian model. Turner and Pearce speculate that existence values may encompass nonutilitarian values and later suggest that existence values are the means whereby individuals reveal their concern for nonhuman intrinsic values.[38] Pearce, Markandya, and Barbier boldly claim "INTRINSIC VALUE = EXISTENCE VALUE" and thereafter only discuss the latter, which they explain can be measured by the contingent valuation method although they note that existence values may reflect some judgment about the rights of nonhuman beings.[39] Thus, the nonhuman intrinsic value concept has become serious enough for neoclassical economists to try to adopt, creating this rather confused literature.

Of course, a nonutilitarian account cannot, by definition, be squeezed into a utilitarian model. Individuals willing to pay for species preservation or biodiversity may be showing respect for the nonhuman intrinsic value they recognize but this trade price fails to measure the intrinsic value. In the same way, paying to help save a person's life fails to reflect the value of his or her life. The reflection of intrinsic values as rights would result in what neoclassical economists term lexicographic preferences, or noncompensatory choices. In this case, intrinsic values in nonhuman animals, plants, or ecosystems are recognized by individuals as a serious constraint on economic trade-offs. Recent studies reveal that approximately 25 percent of respondents to contingent valuation surveys of biodiversity and wildlife show rights-based beliefs by *refusing to bid anything*.[40]

How to Maintain Natural Capital

We have identified two core reasons for wanting to designate a class of inputs to production as natural capital: (1) criticality, and (2) intrinsic value. The main approaches to natural capital maintenance can be evaluated in the light of the preceding discussion. Various suggestions have been advanced to deal with the issues arising from resource flow and pollution absorption capacity limitations. These strategies are set out below, moving from standard economic tools through the development of constraints to a systems approach. The emphasis is on the proposed method of assessing natural capital loss and so identifying the stock to be maintained rather than on the instruments for actual maintenance (i.e., legal, social, economic enforcement measures).

Maintaining Total Economic Value

Some economists have suggested that the value of the environment is summed up in the concept of a "total economic value."[41] In order to assess how far the natural capital stock has been inefficiently used, the value of natural capital is assessed using environmental cost-benefit analysis. A market failure is then corrected by decreasing or increasing what is regarded as either the overuse or underuse of natural capital. Typically environmental goods and services would be undervalued and resource use would as a result be excessive. This approach is normally advised for use at the project level where small changes in development are threatening elements of natural capital. The method essentially aims to assess the value of the income flows lost.

Problems arise due to the limited ability of human preferences expressed in a marketplace to encompass all the information and values identified in the roles of natural capital. First, the existence of perfect information in economic models avoids the issue of how introducing information affects preferences as opposed to merely informing decisions. Second, there is no evidence that human preferences expressed through the market place relate to the relative criticality (as opposed to relative economic scarcity) of forms of natural capital. People appear to value their environmental status quo independent of whether this is linked to a diverse, stable, or robust ecology, and have objected to the removal of slag heaps and the reforestation of denuded hillsides. Similarly, a higher priority is often given to environmental change that happens quickly rather than slowly, probably because rapid change is more psychologically salient (although long-term environmental change may be more serious). Furthermore, individuals tend to place higher economic values on organisms at the tops of food chains, although species at the bottom of food chains are usually much more ecologically significant (donating to save pandas, rather than bamboo). Thus, anthropomorphic values as reflected in the marketplace diverge from ecological values.

More generally, reliance on this idea of assessing the value of the damage associated with natural capital depletion requires that the features of capital lost are commodities, the consumerist approach.[42] If nature fails to fit into this framework, it is excluded from being valued despite its utility. Maintenance of value can be consistent with depleting a resource, or driving a species to near extinction; demand theory suggests a reduced quantity will increase price. For example, speculators are currently hoarding white rhino horns and driving the resource to extinction in order to raise the value of their holdings. The valuations are also entirely anthropocentric and so deny the existence of intrinsic values in nature. Thus, natural capital would be physically depletable.

Compensating Projects

The compensating projects approach, proposed by Klaasen and Botterweg, relies on an adaptation of the free-market-system approach by establishing shadow projects that would compensate for the physical loss of natural capital in physical terms.[43] This technique concentrates on natural capital loss at the level of projects or groups of projects. A developer, whether a government agency or private individual, would provide a project that enhanced environmental functions as compensation for those lost as a result of the development project. Daly has argued for "quasi-sustainable" use of nonrenewables by requiring that any investment in the exploitation of a nonrenewable resource must be matched by a compensating investment in a renewable substitute.[44]

Pearce and associates also support a variant of the compensating projects idea: in this case, by restoration and rehabilitation of the environment. However, they believe this would be too restrictive on a project-by-project basis: "we cannot require each tree, each piece of lost soil, each fine view to be restored." They therefore suggest a "portfolio" approach where the sum of damages due to development projects is balanced by "deliberate creation and augmentation of environmental capital."[45] They would not measure the success of compensation through a cost-benefit analysis, and thus accept an exogenous constraint upon the economic system. However, as discussed in the previous section, the (apparently nonutilitarian) justification for external constraints is left unclear by these authors.

More generally, the aim in adopting a shadow project approach seems to be to avoid the monetary valuation issues. This could in theory include a recognition of the rights of, say, species or ecosystems. For example, species rights might be reflected in the destruction of one habitat requiring the construction of another elsewhere. Yet, similar problems to the monetary valuation approach remain. Exactly what is adequate compensation and how is it to be measured? The portfolio approach assumes that various ecosystem features can be aggregated together, and then humans can "create" natural capital to match at some other group of sites. A project-level approach is therefore preferable in order to avoid such issues. However, this still begs the question as to what compensation humans can make for ecosystem destruction unless the ecosystem is identically reproduced somewhere else on "unused" land. The shadow projects themselves will change the stock of natural capital. For example, if a meadow were to be flooded to create a replacement for a wetland area destroyed by development, the stock of natural capital is reduced by one meadow (after the compensating project). Only if the land on which the shadow project is to be "created" is ecologically worthless can the stock of natural capital be held constant by a compensating project; replacement ecosystems would be inferior in terms of their structural development, and mechanistic replacement can ignore the ecological characteristics of a site. Also, as Munro and Hanley point out, planning agencies with multiple goals might trade off various project features so that natural capital would be reduced by shadow projects.[46]

Scientific Thresholds

A third approach would be to allow a "scientific" determination of the limits to be set on the use of natural capital. This views the world as a complex machine that needs close analysis by atomistic reductionism to identify all the working parts and how they fit together. The maintenance of natural capital would then be achieved by defining a critical natural capital stock, which at the limit would include all natural capital, to be preserved absolutely. Victor has argued in favor of a set of biophysical constraints on an economy that would define the conditions of sustainability in direct biophysical units.[47] Thus, the requirements of the ecosystem are to be taken into account in so far as they support the sustainability of the economic system. Common and Perrings define an ecological-economic approach with resources allocated so as to be consistent with the protection of stability in both the system as a whole and key components of the system.[48] Their ecological sustainability is a physical concept based upon population indicators (for stability) and indicators of responsiveness of systems parameters to perturbations in resource stocks (for resilience).

The levels to which scientific reductionism can assess ecosystem constraints is limited by the current level of human knowledge and ultimately irreducible ignorance.[49] Thus, gaining a consensus on the appropriate thresholds is far from discovering an "objective" truth. As science tends to learn by doing, there is also some risk of awaiting the evidence, and getting confirmation of critical natural capital loss once it has become irreversible, or the system has started an irrevocable transition. However, some aspect of this approach will be necessary in the maintenance of natural capital.

The scientific threshold approach could be used to acknowledge nonhuman intrinsic values and respect them. Scientific research into system boundaries could include the impacts on nonanthropocentric values and set limits that allowed for a wide range of necessary conditions. This would require a recognition by society of the values that scientists are to serve. Currently the drive toward market economies negates concerns for either intrinsic values or ecological stability even when the regulatory authority may recognize their importance.[50]

In theory, this scientific approach is operated by the United States in its use of primary air pollution standards, which aim to protect human health regardless of cost. However, the standards have had enforcement problems with some regions consistently failing to meet the requirements. This is worrying when the scientific evidence is in favor of tightening standards even further. For example, tropospheric ozone (e.g., Los Angeles smogs) was restricted to protect human health, but the old, asthmatic, and very young are still susceptible to harm. In addition, the initial lack of data when designating a threshold means long-term effects tend to be ignored. Long-term exposure to relatively low doses of ozone can shorten life and affect lung development. These problems require a tightening of the standard, but as the standard is already exceeded, the problem of how to achieve the new goal arises. Some economists would then argue that the objective scientific approach be replaced by cost-benefit analysis as the appropriate decision-making tool to decide the acceptable probability of failure to meet a given standard, the margin of safety.[51] Others might regard the issue more widely and desire a change in human behavior, in this case new approaches to transportation. The point is that scientific information can easily require lifestyle affirmation or change but the decision over the lifestyle adopted is outside of the scientists' remit. Thus, Common and Perrings argue for overriding consumer sovereignty where it supports ecologically unsustainable preferences and technologies, but they fail to address the issue of who does the overriding or how.[52]

The enforcement of scientific thresholds raises many issues on the role of scientific judgment in the decision-making process (issues that are also relevant to the policy prescriptions of positive economics). Scientific evidence on environmental problems is normally a prerequisite for action. However, this is very different from allowing politically appointed oligarchies of "experts" to decide upon the appropriate levels of natural capital. The dangers of oligarchies in action can be seen by another example from the United States; appeals against the Endangered Species Act lead to the creation of a politically appointed quango (board) of nonexperts with the remit to decide whether a development project is actually a threat to a species. The seven-person Endangered Species Committee has been nicknamed the "God Squad," due to their power of life and death in cases such as the Mount Graham red squirrel and the northern spotted owl.[53] Replacing the committee with expert scientists leaves the political and economic issues unresolved. For example, part of the issue here concerns the perception of risk to a particular species. The general public has been observed to reject very low-probability, high-loss risks that experts judge to be acceptable.[54] Thus, the experts could vastly underestimate the potential welfare costs that these risks impose upon people.

The argument here is about whose probabilities of uncertain future events count, those of the experts or the public, and whose preferences count, those of the select committee or individuals who face the risk/loss. This suggests that the scientific priorities should be set by the wider community, rather than just the scientific community. For example, biodiversity protection has been approached from two directions, one aimed at species preservation (as in the United States) and the other at ecosystems conservation (as in the United Kingdom). Hence, the problem of the God Squad has been avoided in the United Kingdom (although there are other problems).[55] Science is itself at least in part a social activity, and its priorities are influenced by political and economic factors. Outside of the realm of the "enlightened philosopher king," science cannot provide constraints on behavior based on pure, abstract, and absolute truth. Furthermore, even given a scientific consensus as to the optimal target for some particular use of natural capital, there will be a number of political and economic choices to be made as to how to achieve that target. Thus, intrinsic, nonhuman values and duties to future generations take on a role in constraining scientists similar to the one ecology has been argued to take in constraining economics. These issues are of high priority given the role of science in environmental policy formation.

Systems Approach

At the opposite extreme to atomistic reductionism is the attempt to consider the whole system in a holistic manner. Walters describes how resource management can benefit from systems modeling even when the models themselves are never actually employed.[56] The process of conceptualizing systems boundaries in an interdisciplinary setting is useful in and of itself. He also emphasizes the need to address uncertainty and the role of adaptive management rather than optimal solutions. Extended systems approaches have been suggested by Berkes and Folke and developed by Clayton and Radcliffe.[57]

The vital task in the development of strategies for the sustainable management of natural capital is to integrate the critical scientific, socioeconomic, and philosophical perspectives. Scientific research allows the development of understanding of the behavior of the biological, ecological, geochemical, and other processes that shape the global environment: allowing society to monitor change, identify trends, and predict possible outcomes. However, all such knowledge is probabilistic, and all decisions must be made in light of risk, uncertainty, and ignorance. Thus, there will always be political and economic choices to be made in meeting any given scientifically determined target. Socioeconomic analysis is essential if we are to develop techniques for assigning and incorporating environmental values, where appropriate, into economic and related decision making. This allows the choice of economic tools to achieve desired ends with the minimum of means and adverse effects. Other nonefficiency goals are also of concern here, for example, redistributive consequences. Philosophical analysis reveals the mix of assumptions that underlie human decision-making processes. In this way, long-term and diffuse relationships between actions and consequences can be brought into a practical ethical framework for decision making.

In practice, therefore, environmental questions are inextricably interlinked with social, economic, and cultural values. Economic systems determine the rate and route of flows of energy and resources from the environment into patterns of human use, and the rate and route of flows of waste and energy and materials from human economic operations back into the environment. These economic systems are, in turn, imbued by cultural values, and underpinned by social and psychological models that influence the way in which people understand their options and make choices. The need is to have some way of incorporating information from such different domains into a single decision-making process.

One approach is to try to map all the relevant information onto one domain. This underlies the attempt in environmental economics, for example, to assign values to ecological and social phenomena, so that they can be brought into cost-benefit analysis. There are three critical problems with this approach. First, information may be lost as knowledge from other domains is translated and mapped. Second, the dynamic interaction of complex ecological and economic systems is neglected and cannot be understood. Third, the way in which information is used, the relationship between information and power, makes the methodology chosen for the assignation of values and weightings highly significant, but this is usually invisible by the time the data have been processed.

Systems theory can offer a multidimensional framework in which information from different domains can be integrated without being forced into one-dimensional mapping. This integration of diverse information can be achieved in various ways. One way is to model the systems concerned in an attempt to draw all key dimensions present into a single model. This is, in general, a highly mathematical approach, and is usually applied to relatively "hard" and quantifiable systems. A more generalist systems approach, which can be extended to "soft" systems that cannot be quantified on an equivalent basis, emphasizes the development of an understanding of the pattern of interaction between the systems concerned. This usually involves drawing on a range of models and analytical tools and constructs developed in the various specialist disciplines involved.

The mathematical and the generalist approaches both require the development of a decision-making process that can accommodate change in a number of nonequivalent dimensions simultaneously. One

Maintenance of Natural Capital 163

way in which this can be done is using multiple indices. Various kinds of graphics can then be used to show movement on these multiple indices, to demonstrate change over time or on a compare-and-contrast basis, and to demonstrate the difference between two or more development options in terms of a complete profile of costs and benefits.

Clayton and Radcliffe employ "Sustainability Assessment Maps" (SAMs) to demonstrate this method (a related but simpler approach was employed by Brendan).[58] A SAM consists of a diagram in which each of the critical dimensions in a complex problem is represented by an axis. Measurement of change or indications of priorities are then mapped onto these axes. The purpose of this approach is to emphasize rather than conceal conflicting values, and to do so in a way that is accessible and intuitive.

When making a development decision, the first step is to identify the critical areas of change (axes of concern), which are key factors in the development process. These could include monetary costs and benefits, profits, number of jobs, types of physical environmental impact, quantity and quality changes in natural capital stock, and so on. The main development options are assessed on the same basis and scored on all of the axes of concern. The scores are then displayed in a SAM graphic. Each option can, as a result, be compared in terms of its overall profile, and in terms of the balance of advantages and disadvantages relative to alternative options. The profiles developed using SAMs for a particular development option can be subtracted from an alternative option to give a new combined SAM that shows the net difference.

In order to help explain how SAMs would operate, we will briefly outline an example using energy policy. The example is for illustrative purposes and therefore may be unrepresentative of actual profiles. Consider a power company deciding whether to install new capacity and facing three alternatives: coal burning, nuclear power, or a tidal barrage. Each option can be considered in turn.

The coal-burning power station will produce large volumes of chemical waste such as carbon dioxide, sulphur oxides, and alkaline ash. The carbon dioxide and sulphur emissions will enter the atmosphere with long-term impacts. The ash might be disposed of in lagoons covering sizable areas near the station. The local and short-term impacts are judged to be within local ecological limits as the chosen site is industrialized, although the lagoons may irreversibly damage local biota. On balance, the risk associated with this option is relatively low. However, the nonlocal and long-term picture is less favorable with contributions to the enhanced greenhouse effect and acidic deposition. This option also incurs upstream environmental costs, for example, mining and transporting coal.

Nuclear power presents a complex range of emissions including radioactive isotopes. The bulk of high-level waste disposal is to be in temporary underground storage. Local and short-term impacts are minimal, which is important because the site is remote and on an environmentally sensitive part of the coast. Long-term storage of high-level waste poses the risk of disaster with a low probability, and passes the problem along to future generations. Upstream costs include uranium mining, fuel processing, and transport of radioactive fuel and waste products.

The tidal barrage has waste products largely limited to the construction phase, unlike the other two options. However, the barrage will flood and destroy the mudflats of the estuary where it is to be built. The estuary is an internationally important site for migratory birds and has a unique endangered plant species. This gives a profile of high site sensitivity and local ecological damage. Yet the long-term and regional/global profile is very favorable with damage limited to a single site.

In addition to these environmental considerations, the construction and operating costs are included. The power company might then proceed with an application to a government agency or during the planning process another option may be identified. Thus, the option of investing in energy conservation could be added to the social decision of how best to use the available resources. These different options are to be analyzed upon several different axes, and twenty-one have been selected for the current analysis. These axes are: critical natural change, other natural change, site value, aesthetic impact, impact scale and risk, emissions (air, land, water, auditory, and electromagnetic and ionizing radiation), net capital growth, application of capital commitment, employment impact, total material and energy input, and resource depletion (fossil fuel, mineral, soils, water, and biotic).

Some examples of the resulting SAMs are shown in figures 7.1, 7.2, and 7.3. There are three concentric rings, which signify, moving from the center outward, regional, national, and global impacts. Each option can be shown individually as is done in figure 7.1 for nuclear power and in figure 7.2 for energy conservation. Next, any two options can be compared (combined) on the same SAM as is done in figure 7.3 for nuclear power versus energy conservation; gray shows a more favorable rating for the energy conservation option and black favors nuclear power.

Maintenance of Natural Capital 165

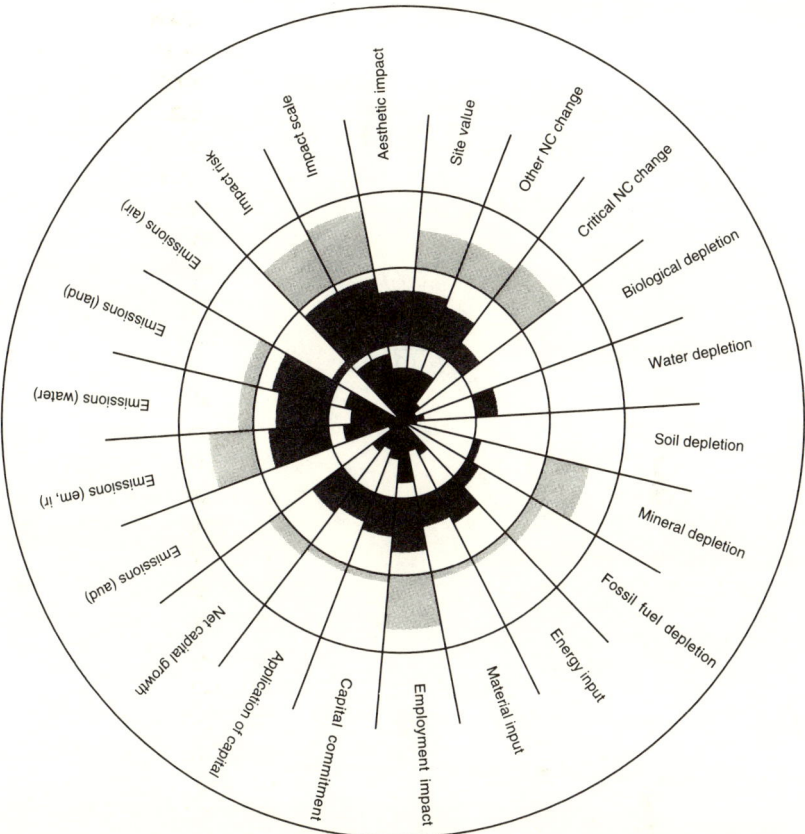

Fig. 7.1. Sustainability Assessment Map for a Proposed Nuclear Power Station. The various sizes of the black marks in the center ring indicate impacts at the regional scale. The gray marks in the next ring indicate impacts at the national scale, and the light gray marks in the outermost ring indicate impacts at the global scale.

The value of the exercise lies in using axes that allow reliable comparisons to be made between options, and to make these choices explicit, so that it is always possible to identify and check assumptions and calculations. SAMs can also be used to clarify areas of disagreement: using the combining process to differentiate scores in two SAMs to create a third, as in figure 7.3. Groups expressing different environmental attitudes can consider a development option and express preferred positions on each axis (e.g., location, scale, water

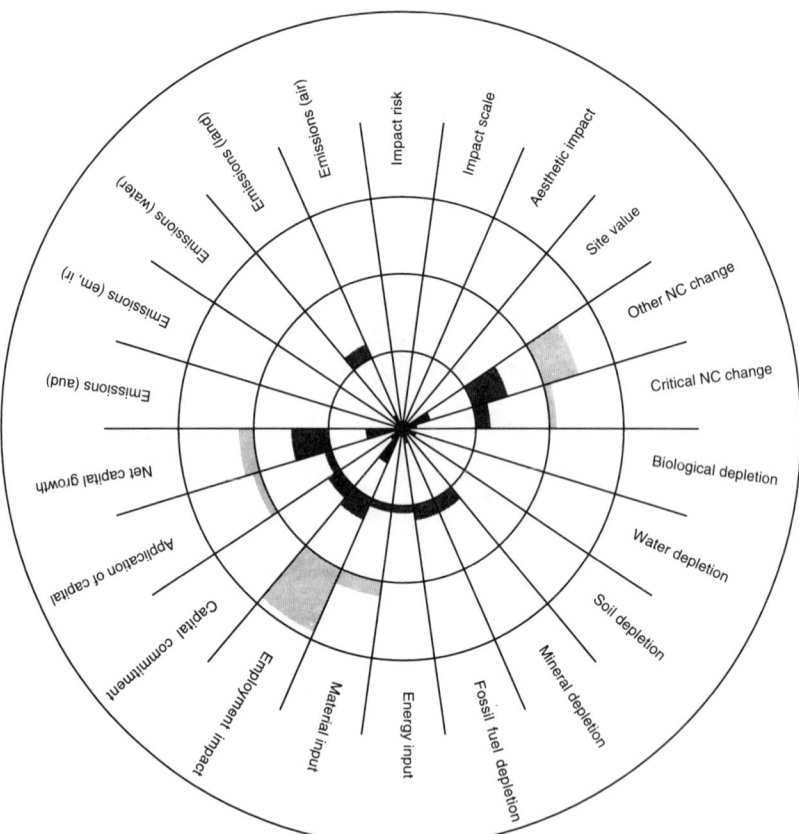

Fig. 7.2. Sustainability Assessment Map for a Proposed Energy Conservation Scheme. The various sizes of the black marks in the center ring indicate impacts at the regional scale. The gray marks in the next ring indicate impacts at the national scale, and the light gray marks in the outermost ring indicate impacts at the global scale.

pollution). The sets of final patterns from the combined SAM will clearly identify areas and degrees of disagreement. This would allow the recognition of a particular position on a particular axis as being "inviolable." Other axes may be seen as negotiable within limits or of little concern in terms of conflict resolution.

The use of concentric overlays in plotting SAMs can show a range of aspects due to a development project on a single diagram. For example, geographic distribution of impacts could be shown within

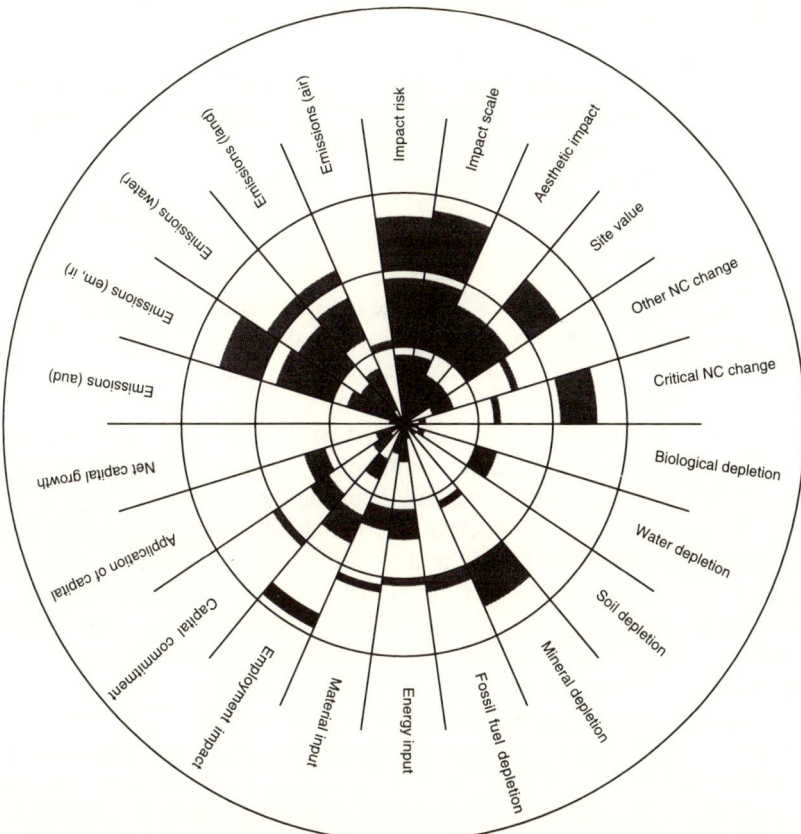

Fig. 7.3. Sustainability Assessment Map Differentiating the Nuclear and Energy Conservation Options. The three concentric rings indicate impacts at the regional, national, and global scales. Gray marks indicate lower impact for the energy conservation option; black marks indicate lower impact for the nuclear power option.

concentric overlays representing local, regional, and global effects, as in the figures presented here. Similarly, intragenerational versus intergenerational impacts can be shown by differentiating each ring by discrete time periods.

In the final instance, where a yes/no decision has to be made, the information must be collapsed into a single dimension. At that stage, SAMs have no advantage over environmental cost-benefit analysis. However, the advantage of SAMs over cost-benefit analysis lies in their function as decision-making aids. The purpose of SAMs is to: make trade-offs more explicit, incorporate incommensurable values, increase accessibility to the decision-making process, encourage identification of a full range of options, and enable effective monitoring of the wider effects of decisions over time.

Conclusions

The justification for wanting to define a set of inputs to production as natural capital seems to require the recognition of values that fall outside of the neoclassical economic model, and therefore provide constraints upon that model. One set of values derives from human ignorance of ecosystems and the need to protect their stability and resilience. The general public is unable to place a willingness-to-pay value on aspects of ecosystems and their functions, which even experts find difficult to explain to each other. The drive toward utility maximization in these circumstances is due to the desires of public agencies rather than a reflection of how individuals actually operate. Individuals confronted by complex problems are likely to refuse to make trade-offs that reduce their ability to adapt to future circumstances, rather than make an irreversible decision based on maximizing current economic gains while facing information limited to their current context. This is a testable hypothesis for which some evidence has been cited.

The second set of values appeals to human acceptance of a wider range of moral agents than themselves. The recognition of nonhuman intrinsic values provides reasons for preventing economic exploitation of natural capital. Under the utilitarian philosophy there can never be absolute or permanent protection. If the arguments of those favoring the existence of intrinsic values in nature are adopted, such protection is justified, and natural capital would be excluded from economic calculations. The problem of maintaining natural capital is then altered into identifying and protecting natural objects and species on grounds of what Hargrove calls intrinsic beauty and interest.[59]

These external constraints leave economic systems facing limits that still need to be defined and accepted by human society. In looking at the approaches to forming and maintaining these constraints in terms of natural capital maintenance, we have argued against narrow views of natural capital in terms of either monetary value or physical characteristics. Such definitions require trade-offs that rely upon questionable assumptions about human abilities to recreate, understand, and manipulate natural systems. The gaps in human knowledge, and our apparent inability to ever fill those gaps, make the maintenance of the fundamental ecosystems and their diversity a necessary step to achieving the relative stability that is implied by sustainability. Thus, we argue in favor of a pluralistic approach that can incorporate a variety of information.

This seemingly leaves the formation of constraints in the hands of experts. Yet the dangers of this approach have also been outlined. This drives us toward an appeal for serious consideration of new institutional structures. As a result, perhaps this chapter is too much in the vein of current nihilistic thinking on environmental valuation, and the extended systems approach, as offered by SAMs, currently lacks the rigor of the alternatives. (Although, pure systems modeling is itself generally a highly mathematical and rigorous discipline.) However, we are trying to move into a more inclusive debate where different disciplines communicate and the benefits of alternative viewpoints are recognized. Neither the free-market economic system nor the expert scientific oligarchies can be relied upon to decide the basis for natural capital maintenance. In a world economic system totally oriented toward making trade-offs, any attempt to suggest limits to trading must either appeal to an alternative view and definition of natural capital (such as nonanthropocentric values), or recognize the need to fundamentally change the way in which humanity perceives development. Both realizations suggest the need for the consideration of multiple values and perspectives, which we suggest an extended systems approach could provide. Underlying this approach is a concern for a more inclusive and holistic view of natural capital maintenance, without which the concept itself becomes meaningless.

Notes

1. M. Common and C. Perrings, "Towards an Ecological Economics of Sustainability," *Ecological Economics* 6 (1992): 7–34.
2. On the philosophical motivation of environmental concern, see P. Craig, H. Glasser, and W. Kempton, "Ethics and Values in Environmental Policy," *Environmental Values* 2, no.2 (1993): 137–157; C. L. Spash, "Ethics and Environmental Attitudes: With Implications for Contingent Valuation," Unpublished manuscript, 1994, 28pp. On the philosophical motivation of environmental economists, see A. Holland, "Natural Capital," in *Philosophy and the Natural Environment,* ed. R Attfield and A Belsey (Cambridge: Cambridge University Press, 1994).

3. M. Redclift, "Sustainable Development: Needs, Values, Rights," *Environmental Values* 2, no.1 (Spring 1993): 3–20, esp. 19.

4. D. H. Meadows, D. L. Meadows, and J. Randers, *Beyond the Limits: Global Collapse or a Sustainable Future* (London: Earthscan, 1992).

5. P. A. Victor, "Indicators of Sustainable Development: Some Lessons from Capital Theory," *Ecological Economics* 4 (1991): 191–213, esp. 196.

6. H. E. Daly, "Elements of Environmental Macroeconomics," in *Ecological Economics: The Science and Management of Sustainability,* ed. R Costanza (New York: Columbia University Press, 1991).

7. F. Berkes and C. Folke, "A Systems Perspective on the Interrelations Between Natural Human-Made and Cultural Capital," *Ecological Economics* 5, no.1 (1992): 1–8, esp. 2; T. Tietenberg, *Environmental and Natural Resource Economics,* 3rd edition (New York: Harper Collins, 1992), 582; M. Jacobs, *The Green Economy: Environment, Sustainable Development and the Politics of the Future* (London: Pluto Press, 1991), 224; C. Clark, "Economic Biases Against Sustainable Development," in *Ecological Economics,* ed. R Costanza, 329; D. Pearce, A. Markandya and E. Barbier, *Blueprint for a Blue Economy* (London: Earthscan, 1989), 3.

8. R. Costanza, H. E. Daly and J. A. Bartholomew, "Goals, Agenda and Policy Recommendations for Ecological Economics," in *Ecological Economics,* ed. R Costanza, 8.

9. K. E. Boulding, "The Economics of the Coming Spaceship Earth," in *Environmental Quality in a Growing Economy,* ed. H Jarrett (Baltimore: John Hopkins University Press, 1966).

10. E. S. Serafy, "The Environment as Capital," in *Ecological Economics,* ed. R Costanza, 175.

11. C. Clark, "Economic Biases Against Sustainable Development," 329.

12. J. M. Hartwick, "Intergenerational Equity and the Investing of Rents from Exhaustible Resources," *American Economic Review* 66 (1977): 972–974.

13. R. M. Solow,"Intergenerational Equity and Exhaustible Resources," *Review of Economic Studies, Symposium on the Economics of Exhaustible Resources* (1974): 29–46.

14. Victor, "Indicators of Sustainable Development."

15. Common and Perrings, "Towards an Ecological Economics of Sustainability," 30.

16. M. Blaug, *The Methodology of Economics* (Cambridge: University Press, 1980), 202–8.

17. As discussed by Victor, "Indicators of Sustainable Development."

18. W. S. Jevons, *The Coal Question: An Inquiry Concerning the Progress of the Nation and the Probable Exhaustion of Our Coal-Mines* (London: Macmillan, 1909).

19. A. Clayton and N. Radcliffe, "Sustainability: A Systems Approach; A Technical Report for WWF," Institute for Policy Analysis and Development, Edinburgh (forthcoming).

20. J. H. Lehr, *Rational Readings on Environmental Concerns* (Van Nostrand Reinhold, 1992).
21. J. M. Keynes, *A Treatise on Probability* (London: Macmillan, [1921] 1973).
22. G. Chichilnisky and G. Heal, "Global Environmental Risk," *Journal of Economic Perspectives* 7, no.4 (1993): 65–86.
23. M. Faber, R. Manstetten, and J. L. R. Proops, "Humankind and the Environment," *Environmental Ethics* 1, no.3 (1992): 217–241.
24. A. Stirling, "Confronting Complexity, Ignorance and Discord in Technology Choice for Sustainable Development," Unpublished paper, University of Sussex, 1994.
25. C. Walters, *Adaptive Management of Renewable Resources* (New York: Macmillan, 1986).
26. Foy, G. "Economic Sustainability and the Preservation of Environmental Assets," *Environmental Management* 14, no.6 (1990): 771–778, esp. 772.
27. C. L. Spash and I. A. Simpson, "Protecting Sites of Special Scientific Interest: Intrinsic and Utilitarian Values," *Journal of Environmental Management* 38 (1993): 213–227.
28. J. Passmore, *Man's Responsibility for Nature: Ecological Problems and Western Traditions* (London: Duckworth, 1974).
29. J. Bentham, *An Introduction to the Principles of Morals and Legislation* (Oxford: Athlone University Press, [1789] 1970).
30. A. Leopold, *A Sand County Almanac*, introduction by R Finch (Oxford: Oxford University Press, [1949] 1987), 224–225.
31. J. Rodman, "Four Forms of Ecological Consciousness Reconsidered," in *Ethics and the Environment*, ed. D Scherer and T Attig (Englewood Cliffs, NJ: Prentice-Hall, 1983).
32. R. Attfield, "The Good of Trees," in *People, Penguins and Plastic Trees*, ed. D. Van De Veer and C. Pierce (Belmont, California: Wadsworth, 1986).
33. Holland, "Natural Capital"; Pearce et al., *Blueprint for a Blue Economy*.
34. Common and Perrings, "Towards an Ecological Economics of Sustainability."
35. R. K. Turner and D. W. Pearce, "The Ethical Foundations of Sustainable Economic Development," London Environmental Economics Centre paper 90–01, 1990, 40 pp.
36. Turner and Pearce, "The Ethical Foundations of Sustainable Economic Development," 31–32.
37. Turner and Pearce, "The Ethical Foundations of Sustainable Economic Development," 34.
38. Turner and Pearce, "The Ethical Foundations of Sustainable Economic Development," 16, 29.
39. Pearce, et al., *Blueprint for a Blue Economy*, 62, 77.

40. C. L. Spash and N. Hanley, "Preferences, Information and Biodiversity Preservation," *Ecological Economics* 12 (1995): 191–208; T. H. Stevens, J. Echeverria, R. J. Glass, T. Hager and T. A. More, "Measuring the Existence Value of Wildlife: What do CVM Estimates Really Show?" *Land Economics* 67, no.4 (1991): 390–400.

41. A. M. Freeman, *The Measurement of Environmental and Resource Values: Theory and Methods* (Washington, DC: Resources for the Future 1993), 147; D. Pearce, E. Barbier and A. Markandya, *Sustainable Development* (London: Earthscan, 1990); D. Pearce and K. Turner, *Economics of Natural Resources and the Environment* (Hemel Hempstead: Harvester Wheatsheaf, 1990).

42. See C. L. Spash, "The Political Economy of Nature," *Review of Political Economy* 7, no.3 (1995): 293-307.

43. L. Klaasen and T. H. Botterweg, "Project Evaluation and Intangible Effects: A Shadow Project Approach," in *Environmental Economics, Volume 1*, ed. P. Nijkamp (Leiden: Martinus Nijhoff, 1976).

44. H. E. Daly, "Toward Some Operational Principles of Sustainable Development," *Ecological Economics* 2, no.1 (1990): 1–6.

45. Pearce and Turner, *Economics of Natural Resources and the Environment*, 225.

46. A. Munro and N. Hanley, "Shadow Projects and the Stock of Natural Capital: A Cautionary Note," Economics Discussion Paper, University of Stirling, 1991.

47. Victor, "Indicators of Sustainable Development."

48. Common and Perrings, "Towards an Ecological Economics of Sustainability," 30–31.

49. Faber et al., "Humankind and the Environment."

50. Spash and Simpson, "Protecting Sites of Special Scientific Interest."

51. A. J. Krupnick, "Economics and the ozone standard, " *Resources* 92 (Summer 1988): 9–12; Freeman, *The Measurement of Environmental and Resource Values*, 266.

52. Common and Perrings, "Towards an Ecological Economics of Sustainability."

53. National Wildlife Federation, "'God Squad' Targets the Owl," *Enviro-Action* 9. no.9 (1991): 6–8.

54. Freeman, *The Measurement of Environmental and Resource Values*, 260.

55. See, for example, Spash and Simpson, "Protecting Sites of Special Scientific Interest."

56. Walters, *Adaptive Management of Renewable Resources*.

57. F. Berkes and C. Folke, "A Systems Perspective on the Interrelations Between Natural Human-Made and Cultural Capital"; Clayton and Radcliffe, "Sustainability: A Systems Approach."

58. Clayton and Radcliffe, "Sustainability: A Systems Approach"; C. C.

Brendan, "A Systems Management Approach to Evaluating Sustainability in the Context of Tourism: with Particular Reference to the Badenoch and Strathspey District of the Scottish Highlands," Unpublished M.Sc. Dissertation, University of Edinburgh, 1992.

59. E. C. Hargrove, *Foundations of Environmental Ethics* (Englewood Cliffs, NJ: Prentice-Hall, 1989).

Wilderness Management

Roger Paden

In this essay, I will discuss several different kinds of what could be called "natural areas policies," that is, policies that might be adopted to regulate our treatment of, and interaction with, large natural areas, such as parks, forests, grasslands, lakes, seashores, and marine habitats; in short, actual or potential "wildernesses." Each of these policies rests on a complex set of moral and conceptual presuppositions. In what follows, I will attempt to articulate some of the presuppositions of each of these policies in order to assess their strengths and weaknesses.

Although I am particularly attracted to one of these policies, which I will call a "process preservation policy," I also find it to be the most problematic. As a result, much of my discussion of it will focus on uncovering some of the conceptual problems that its partisans must face. The purpose of this essay, therefore, is not to provide this policy with a strong philosophical foundation. Instead, I only intend to examine the problems that await anyone who undertakes this task. My goal, that is, is clarity of exposition, not certainty of argument.

I believe that the best way to approach the task of characterizing these policies is through an examination of the history of natural areas policy in the United States. Therefore, I will develop my ideas through a very brief, highly episodic, and somewhat tendentious discussion of the history of the policies that the federal government has adopted with respect to natural areas.

I

The first, and longest lasting, federal natural areas policy, was one of simple and rapid exploitation. For nearly two hundred years, the federal government has sought to aid private developers in exploiting nature (including those natural areas owned by the government) in hopes of encouraging rapid economic expansion. Driven by a belief in an American "Manifest Destiny," as well as by a desire to reap the social, economic, and political advantages that were thought to follow from a policy of rapid expansion and development, the federal government sought, during this period, to encourage development through a policy granting title to sizable blocks of western land to anyone promising to settle and/or develop it. The high water mark of this kind of policy occurred, arguably, in 1862, when the Congress passed the Homestead Act, which, in effect, made Locke's theory of property ownership through development, which he referred to as "acquisition through labor," the law of the land.

Underlying this development policy were two characteristically Lockean assumptions, that nature was virtually inexhaustible and that it was morally insignificant. The first assumption was taken directly from Locke: In the "vast wilderness" and "vacant places of America," there is enough land for all who are but willing to invest their labor into it.[1] The second follows from what has been termed an "anthropocentric ethic," which is implicit in Locke's view of the value of natural objects. An anthropocentric ethic—in contrast to a "biocentric ethic," which grants intrinsic value to at least some natural objects—holds that things are valuable only to the degree to which they are useful in satisfying human interests, which alone have intrinsic moral value.[2] Therefore, in the anthropocentric view, nature can have only an instrumental value. Given such an ethic, together with a belief in nature's inexhaustibility, it follows that the correct natural-areas policy would be to make use of these areas as quickly as possible to satisfy as many human interests as possible. Such a "development policy," that is, follows naturally from the conjunction of these two ideas.

With the closing of the American frontier—sometime during the 1890s if we use Frederick Jackson Turner as our guide—and the increasingly obvious "overall decline in agriculture in the eastern United States after 1830," due in large part to soil depletion and erosion, anthropocentric arguments in favor of another kind of policy, a "resource conservation policy," appeared to become more persuasive.[3] This policy was based on the idea that natural resources are limited and that, therefore, they should be wisely husbanded or "conserved" so as to insure their best use. As Gifford Pinchot, the first Chief of the U.S. Forest Service and Teddy Roosevelt's environmental adviser, noted, prior to the adoption of this policy, "the nation was obsessed . . . by the fury of development. The American Colossus was fiercely intent on appropriating and exploiting the riches of the richest of all continents—grasping with both hands, reaping where he had not sown, wasting what he thought would last forever.[4] However, Pinchot argued, with the realization that natural resources are limited, comes the recognition that the policy of rapid development is unsound.

As a result, the federal government, under the influence of the Progressive movement, began to establish a large number of conservation programs, such as the federal soil and water conservation programs that are still with us today.[5] These programs attempted to conserve natural resources, expending them only on projects that are either necessary or promise an especially high rate of return. Ideally, these programs sought to promote the indefinitely sustainable use of those resources. Generally, during the last one hundred years, the federal government has been slowly changing its policy approach from one of promoting rapid development to one of promoting sustainable "resource management" or "conservation." This change, it goes without saying, is not yet complete.

Although it is possible for a conservation policy to be based on either anthropocentric or biocentric beliefs, usually conservationists have grounded their policies on a purely anthropocentric ethic. Thus, many early conservationists, such as Pinchot, have publicly argued that, although natural areas are intrinsically valueless, because our needs for natural resources are both great and growing, and because the supply of those resources, both relative to those needs and absolutely, is shrinking, those resources will come to have an increasingly high "use" or instrumental value, and therefore, they should be protected and conserved.

Thus, the change from a development policy to a conservation policy was, generally speaking, a change in policy only; it did not represent a change in underlying moral positions: It was a change in means, but not in ends. Throughout, the federal government continued to treat nature as an exploitable economic resource. Throughout, the government, in effect, remained committed to an anthropocentric approach to nature. Both the development policy and the conservation policy, that is to say, were defended on the basis of an anthropocentric ethic. Where the partisans of the two policies disagreed was in their views on the abundance of nature.

All this is easily understandable. A rational anthropocentrist who believed natural resources to be infinite or, at least, extraordinarily abundant would naturally adopt a policy of rapid development in order to maximize human values. However, if she believed that natural resources were scarce and that their rapid exploitation would lead to their eventual exhaustion, she would naturally adopt a conservation policy to achieve that same end. Thus, the change from a development policy to a conservation policy was due primarily to a change in the federal government's conception of nature; to a change, that is, in its view on the relative abundance of our natural resources.

At the same time that the federal government was beginning to implement various conservation programs, it inaugurated another program, which has grown into today's National Parks Service. This program, in the contemporary popular mind at least, best represents a third general kind of natural areas policy, a "preservation policy."[6] However, while the national parks seem to symbolize the idea of the permanent preservation of intrinsically valuable natural areas, they were not, at first, given such a preservationist justification. Instead, as Eugene Hargrove in his book, *Foundations of Environmental Ethics*, has indicated, in the congressional debates concerning the establishment of both Yellowstone and Yosemite, the parks were originally justified on conservationist grounds.[7]

According to Hargrove, the argument that seemed to carry the day for the parks was a bad-faith conservationist argument that contended that the areas in question should be preserved because they were so desolate, inhospitable, and distant that they lacked economic or instrumental value.[8] Preserving them, this argument urged, was the best way to "use" them. While this claim might have been politically persuasive in a Congress that, only ten years earlier, had passed the Homestead Act, it is so transparently bad, that it is best understood as a cover for more straightforward, but politically unpopular, preservationist arguments.

Such cover was needed, for, generally speaking, not only do preservationist policies greatly differ from developmental policies, but they also *seem* to differ from conservationist policies as well. For example, while conservationists argue that we should protect natural areas from excessive economic development, preservationists seem to argue that these areas should be protected from all development. While conservationists argue that we should protect nature for our future benefit, preservationists seem to argue that we should protect nature permanently, regardless of future benefits. Finally, while conservationists typically use anthropocentric arguments, preservationists often use biocentric arguments.

However, these apparent differences may be deceiving. Preservationists, as in the Yellowstone case, often do make economic and anthropocentric arguments for their position. Sometimes, unlike those made for the preservation of Yellowstone, they are even economically sound, based, for example, on the idea that natural areas can be harmlessly run as profitable tourist attractions. Of course, it is important to note that the economic uses endorsed by preservationists, unlike those favored by conservationists, are nonconsumptive and therefore sustainable (although perhaps not as profitable in the short run). Moreover, preservationists often make another kind of (generally anthropocentric) argument based on a position that Hargrove has characterized as "therapeutic nihilism."

This position is based on three assumptions. First, it is assumed that nature has only instrumental value. Second, it is also assumed that nature is so complex that, given our limited knowledge and abilities, we cannot predict the ultimate effects that our interference with nature might have on it. As a result, any interference, even the best intended and informed interference, will very likely have unintended—and often negative—side-effects. Given the third assumption, that we are absolutely dependent on nature, it follows that any interference with nature could lead to disastrous results. Therefore, to complete the argument, we should attempt to interfere with nature as little as possible. We should preserve it; and we should "let nature take its course." But we should do this for *our* own good.

This is an interesting argument. Notice that, with a few modifications, it could also be advanced by biocentrists. Most recently, it has been used to urge the advisability of preserving the Amazonian rain forest, which, it is said, may influence the climate in ways that are not yet fully understood. However, as Hargrove points out, given the rapid advance of science, this argument may not have a very long shelf life. Moreover, as we have already interfered with nature on a grand scale, it may be a little late to embark on this policy, as our only hope of preventing an environmental disaster may be to continue to interfere with nature to correct the damage that we have already done.

Preservationists, however, also offer purely biocentric arguments in support of preservationist federal policies. Indeed, this is becoming the favored argument, one that I will take to be characteristic of the preservationist position. Perhaps most famously, that ex-federal land manager and bureaucrat, Aldo Leopold, used biocentric arguments in favor of preservationist policies to argue that we ought to adopt a policy of protecting the "integrity, stability, and beauty" of nature, which Leopold refers to as, "the biotic community."[9] In the remainder of this chapter, I will assume that the preservationist position is essentially biocentric.

II

Although we can differentiate preservationist policies from conservationist policies on the basis of the former's biocentric assumptions that nature is intrinsically valuable, the preservationist position itself is not completely clear due to an ambiguity in a central term, "nature." That term is open to at least two interpretations: It can be understood to refer either to the set of all natural entities (however many we might wish to include in it), or to the long-term historically specific processes that produce those entities. Moreover, these differing interpretations of "nature" support two diametrically opposed kinds of preservation policies, which I will call a "products preservation policy" and a "process preservation policy," respectively. Unfortunately, many environmentalists fail to distinguish between these two kinds of policies. As a result, they are often unable to develop coherent programs or offer consistent justifications of them. Therefore, it is important to understand this distinction. In order to make it clear, it would be helpful to examine two particular preservationist programs directed in part at national parks and wilderness areas.

The first program, representing products preservation policies, is aimed at the preservation of endangered species. In 1973, Congress passed the Endangered Species Act, which gave unprecedented legal protection to whole species of animals. Not only did this law seek to protect endangered species directly by protecting their members from both recreational and commercial killings, but it also sought to protect them indirectly by preserving and protecting their "critical habitats."[10] In 1978, the Supreme Court ruled that the act required that endangered species and their habitats must be protected "without exception" at "whatever the cost," and that their protection "overrides even the primary missions of federal agencies."[11] This interpretation allowed preservationists to use the act as a tool in protecting both endangered species and their wilderness habitats. However, that ruling may have represented the high-water mark in the attempt to use federal law to protect wilderness areas, as it led to a series of amendments that, taken together, significantly weakened federal protection of wilderness areas.[12]

The second preservationist program, representing process preservation policies, is the park service's policy allowing naturally occurring forest fires to burn themselves out as long as they do not endanger high-use areas or famous monuments, or threaten to become uncontrollable. First adopted in 1964, the origins of this fire-management program have been traced as far back as the 1916 National Park Service Organic Act.[13] The program was reviewed in 1989, after the devastating fires in Yellowstone and other western parks during the summer of 1988, and was generally reaffirmed and strengthened.[14] In effect, this program urges a different conception of wilderness preservation, one that seeks to "neutralize the unnatural influences of man, thus permitting the natural environment to be maintained essentially by nature."[15]

Both these programs are preservationist in character. Moreover, both programs are at least arguably biocentric, as they seem to take the preservation of nature to be valuable in itself. On the other hand, the two programs are radically different, not only in the kinds of actions that they require, but also in their conceptual basis.

The most obvious way in which the two programs differ is in their practical implications. A program mandating the preservation of species will often require federal managers to interfere in natural systems. To be sure, most recent extinctions have been caused by human interference with nature, and often these extinctions can be prevented if this interference is simply stopped. However, natural extinctions do occur. If a program seeks to protect species because they have intrinsic value, then it would seem wrong in principle to distinguish between extinctions on the basis of their cause. Moreover, in those cases of natural extinction (and probably in many cases of artificial extinctions, as well), the most effective preservation strategy might require substantial human interference with the environment to preserve the habitat of endangered species artificially. This, however, would run counter to the noninterference in natural processes strategy that underlies the park service's fire-management program.

More important, the two programs seem to be based on different conceptual grounds. In particular, although both programs seek to protect nature, because they understand nature differently, they seek (inconsistently) to protect different aspects of nature. The Endangered

Species Act is based on the idea that we have a direct duty to distinct parts of nature—to natural individuals and/or classes, specifically, in this case, to endangered species, but by implication to all species and their members—to help maintain them and to insure their survival. In effect, this program seems to conceive of nature as consisting of discrete parts or products, each of which is intrinsically valuable. On the other hand, the park service's fire-management program is not aimed at preserving any particular existing part of nature. Instead, it simply seeks to isolate large natural systems from human interference. In effect, it seems to conceive of nature not as a collection of parts, but as a dynamic process.

It is possible to generalize this discussion. The products preservation policy and the process preservation policy can be distinguished on three different grounds. First, the two general policies will often suggest opposing programs. Second, these general policies differ in their conception of nature. Third, I will argue, these policies must also differ in their moral justification.

The most obvious difference between the two policies lies in the kind of programs they support. The process preservation policy requires us to let natural fires burn, pests infest, and diseases run their natural course, even if doing so threatens intrinsically valuable species and/or individuals. On the other hand, the products preservation policy requires us to protect those species, even if doing so necessitates the creation of largely "artificial" habitats.

This divergence in programs is based on a conceptual difference. As argued above, the two policies differ fundamentally in their conception of nature. While one policy understands nature to be a process, the other understands it to be a collection of loosely related products. Unfortunately, many preservationist manifestos fail to maintain this distinction. One example of this failure is Leopold's often-quoted maxim that we ought to seek to maximize the "stability" and "integrity" of the "biotic community." This maxim may be conceptually inconsistent, and it may, therefore, be self-contradictory in practice, depending on how "stability" and "integrity" are understood. Of course, "stability" can mean several things.[16] However, the most natural interpretation of this term would have it that a thing is stable only if it doesn't change, and a thing doesn't change only if its parts remain the same or are replaced with virtually identical parts. On the other hand, "integrity"—which in its native human context refers to the ability to remain faithful to one's principles and, thus, not be easily diverted from one's essential purposes—in this biotic context, is most naturally interpreted as referring to the continued self-determination of nature by natural processes. Nature's integrity is respected, in this interpretation, to the degree that it is not driven by external human interference away from its natural or typical course of development.

The difference between the meanings of these two terms, however, merely reflects a deeper ambiguity in Leopold's formulation. This ambiguity is located in the phrase, "the biotic community." Clearly, this phrase is metaphorical in that it draws our attention to the similarities between the human community and the biotic community. No doubt Leopold wanted to stress the shared characteristics of diversity, interdependence, and egalitarianism in both communities. However, there are at least two distinct ways to understand the "human community." An individualist will understand that community to be a collection of essentially independent, but loosely related, morally considerable persons. A holist, on the other hand, might understand it as a tradition of related institutions, practices, and shared meanings that is evolving through history.[17] Similarly, there are two ways to understand "the biotic community" or "nature": In a "holistic model," it would be a historically situated, evolutionary system in the process of internally driven development (a "process"). In an "individualistic model," it would consist of a large number of relatively independent parts ("products"). The kind of preservation policies we adopt, therefore, will be a function, in part, of our understanding of nature. The holistic understanding, which is consistent with the attempt to protect nature's "integrity," would lead us to adopt a process preservation policy, while the individualistic understanding, which is consistent with the attempt to promote "stability," leads to a products preservation policy.

Failure to distinguish these two views of nature will only lead to confused moral arguments and inconsistent policies. In addition, however, it will result in a confusion as to the moral foundations of preservationist programs. This is the case because the two policies necessarily differ in their moral justification.

The paradigm of moral philosophy that has dominated environmental ethics has been one based on the distinction between instrumental and intrinsic value, in which intrinsic value is understood in terms of the satisfaction of interests. Things with interests, in this view, have intrinsic value; things that can satisfy interests have instrumental value. On the basis of this paradigm, many environmental ethicists have attempted to argue that nature and natural entities have intrinsic value because they have interests.[18] This strategy seems appropriate in the justification of a products preservation policy, *if* it could be shown that some natural beings, for example, animals, have interests and are, therefore, intrinsically valuable and worthy of preservation. I believe that such a strategy is plausible, at least in the case of animals.

However, I believe, such a strategy is inappropriate in the justification of a process preservation policy, because it presupposes that natural processes, such as ecosystems, have interests, when in fact they don't. This last claim is controversial. Many environmentalists have claimed that ecosystems have purposes and, indeed, a great deal of evidence for this claim can be found in the striking stability and resilience of ecosystems. Their ability to maintain a dynamic equilibrium in the face of massive human interference has led some to believe that they have an "interest" in maintaining that goal.[19]

Note the connection, implicit in this last sentence, between the having of interests and goal-directedness. The two are, I believe, conceptually related in that to have an interest is to be (at least potentially) goal directed.[20] Therefore, to argue that ecosystems have interests is to assume that they are goal directed. However, I would argue, despite the fact that they typically maintain a dynamic balance, ecosystems are not goal directed. This is the case because, for a system to be goal directed, it must not only maintain a persistent state by means of behaviors that are sufficiently persistent and plastic, but it must engage in that behavior in order to maintain that state.[21] That is to say, a system is goal directed only if (1) its behavior tends to bring about its "goal" and (2) its behavior occurs because (i.e., is brought about by the fact that) it tends to bring about that goal.[22]

This is a fairly liberal understanding of goal directedness. Using it, it is possible to understand both animals and plants as goal directed. Thus, on this basis, it is possible to attribute interests to them. It is even possible on this basis to attribute goals and interests to robots and guided missiles. However, the stability of an ecosystem is not sufficient evidence upon which to base a claim that it is goal directed. Instead, its stability, clearly, is a result of the goal-directed behavior of the various organisms that make it up. The stability of an ecosystem, that is to say, is a behavioral "by-product," not a "goal."[23] If this is the case, however, it is a mistake to attribute interests to ecosystems.

However, if it is a mistake to attribute interests to an ecosystem, it would be impossible to ground any direct obligation to preserve ecosystems on the idea that we have an obligation to protect the interests of all beings equally. Moreover, if a process preservation policy cannot be based on the general idea of the equal protection of interests, the moral foundations of that policy necessarily will be very different from those of a product preservation policy. While a products preservation policy is typically based on the idea that we should attempt to equally protect the interests of all biological organisms, a process preservation policy cannot be so grounded. These two kinds of preservation policies necessarily rest on very different moral foundations.

III

I have indicated four kinds of natural areas policies: the development policy, the conservation policy, the products preservation policy, and the process preservation policy. Each is uniquely characterized by its conception of nature. In addition, the two preservationist policies can be distinguished from both the development policy and the conservation policy on the basis of their moral foundations. While the characteristic justification of a preservationist policy is biocentric, the typical justification of development and conservation policies are anthropocentric. Finally, the two preservationist policies can be distinguished from each other on the basis of their policy recommendations, their conception of nature, and their moral foundations.

Of the four, the policy with which I am most sympathetic, the process preservation policy, faces the most difficult problems. One of those problems is conceptual. The other is moral. Conceptually, this policy seems to be based on a distinction between natural processes and artificial processes or "human interference." In practice, this distinction is relatively easy to make. In theory, however, it is open to the objection that human beings are merely another part of nature. If so, then this distinction is clearly a spurious one; the residue, it might be added, of an older anthropocentric metaphysics that in other contexts preservationists have rejected.

However, it might be possible to maintain this distinction without reverting to a dualistic metaphysics. Indeed, one way to so maintain this distinction has already been suggested in this paper: We are beings with interests and goals, while nature is not. However, if we do distinguish humans from nature in this way, then we could not consistently claim that our moral duties to nature are based on nature's "interests."

If, in order to defend a process preservation policy, we distinguish humans from nature in this way, then we cannot base the claim that nature is morally significant on the claim that it has interests. This, however, leads to the second problem. If a process preservation policy cannot be based on this position, then we will find a justification for that policy only if we reject the paradigm of moral philosophy that holds that moral duties are essentially related to interest satisfaction. There are several reasons that may lead environmentalists to reject this moral paradigm. Historically, it has been used to justify the kinds of social and economic systems that have caused the greatest degree of environmental damage. Moreover, it presupposes such a "thin" theory of the good, that it has great difficulty accounting for a great variety of relatively strong moral intuitions.[24] However, at least for environmentalists who are interested in justifying a process preservation policy, the best reason to reject this paradigm is that it is incompatible with a process preservation policy.

Already several environmental ethicists, most notably, perhaps, Sagoff and Hargrove, have begun work on a different kind of ethical theory that does not take the satisfaction of interests to be the sole intrinsic value.[25] Interesting enough, both of these philosophers have attempted to develop their ideas through an investigation of our moral traditions and attitudes toward the environment. Clearly, there is some intellectual elegance in the idea that a process preservation policy must be defended by a tradition-oriented ethical theory. However, this should not blind us to the fact that we are not yet in possession of such a theory. Whether such a theory can, in the end, support a process preservation policy remains to be seen. However, if a defense of such a policy can be constructed, I believe that it will be constructed in this way.

Notes

1. John Locke, *The Second Treatise on Government* (Indianapolis, Ind.: Hackett Press, 1980), 22–23.
2. For a discussion of this distinction, see George Sessions and Bill Devall, *Deep Ecology: Living as if Nature Mattered* (Salt Lake, Utah: Peregrine Smith, Inc., 1985), 41–50.
3. Frederick Jackson Turner, *The Frontier in American History* (New York: Holt, Rinehart, and Winston, 1962); Margaret W. Rossiter, *The Emergence of Agricultural Science* (New Haven, Conn.: Yale University Press, 1975), xii.
4. quotation from Gifford Pinchot, *Breaking New Ground* (New York: Macmillan, 1947), 23.

5. Pinchot, *Breaking New Ground*, 109–26.

6. Bryan Norton, "Agricultural Development and Environmental Policy: The Conceptual Issues," *Agriculture and Human Values* 2 (1985): 63–71; and John Passmore, *Man's Responsibility for Nature* (London: Duckworth, 1974), 101–26.

7. Eugene C. Hargrove, *Foundations of Environmental Ethics* (Englewood Cliffs, N.J.: Prentice Hall, 1989), 48–52.

8. Hargrove, *Foundations of Environmental Ethics*, 48–52.

9. Aldo Leopold, *A Sand County Almanac: With Essays on Conservation from Round River* (New York: Ballantine Books, 1966), 262.

10. Roderick Nash, *The Rights of Nature: A History of Environmental Ethics* (Madison: Wisconsin University Press, 1989), 174–78.

11. Holmes Rolston, *Environmental Ethics: Duties to and Values in the Natural World* (Philadelphia: Temple University Press, 1988), 126.

12. Nash, *Rights of Nature*, 174–78. Unfortunately, the current U.S. Congress seems intent on further weakening and perhaps even destroying this act under the guise of "regulatory reform."

13. USDA, *Final Report on Fire Management Policy May 5, 1989*, (Washington, D.C.: U.S. Government Printing Office, 1989), 2.

14. USDA, *Final Report*, 12.

15. A. S. Leopold et al., "Resource Management Policy," in National Park Service, *Compilation of Administrative Policies for the National Parks and National Monuments of Scientific Significance (Natural Area Category)*, rev. ed. (Washington, D.C.: U.S. Government Printing Office, 1970), 106, quoted in Hargrove, *Foundations*, 139–40.

16. Norton, "Agricultural Development," 62–63.

17. Alasdair MacIntyre, *After Virtue* (Notre Dame, Ind: Notre Dame University Press, 1981), 190–209.

18. For example, see Peter Singer, *Animal Liberation*, 2d ed. (New York: Avon Books, 1992), or Rolston, *Environmental Ethics*, for arguments of this sort.

19. For example, Rolston, *Environmental Ethics*.

20. Harley Cahen, "Against the Moral Considerability of Ecosystems," *Environmental Ethics* 10 (1988): 195–216.

21. This is the criterion for goal-directedness suggested by Ernest Nagel, *The Structure of Science* (Indianapolis: Hackett, 1961), 398–421.

22. Larry Wright, "Explanation and Teleology," *Philosophy of Science* 19 (1972): 211–23.

23. Cahen, "Against the Moral Considerability of Nature."

24. For an illustration of this point, see Hargrove, *Foundations*, 168–71.

25. Mark Sagoff, *The Economy of the Earth* (Cambridge: Cambridge University Press, 1988), and Hargrove, *Foundations*.

Mead and Heidegger: Exploring the Ethics and Theory of Space, Place, and the Environment

Eliza Steelwater

The need to theorize relations between human beings and their environment has achieved a centrality in poststructuralist social science that it did not have in modernist inquiry. To the extent that human mastery of the world around us appeared technically feasible, it seemed possible to resolve intellectual, moral, and even political problems in human-environment relations. However, attempts at mastery have unraveled, exposing a host of issues expressed as "uneven development," "marginalization," "sustainability," "environmental ethics," and so on. Current geographical research suggests that the spatiality inherent in such concepts has begun to be understood politically. "Environment," "nature," "wilderness," "resources," "region," "place," "landscape," "relative location" are sites or positions of negotiation, struggle, and domination. More to the point of this paper, the spatial entities themselves are negotiated, their defining characteristics politically arrived at. Is a newly discovered seam of copper a "resource" or a threat? Does locating a toxic waste dump near a low-income residential area "just make sense," or is it a discriminatory act? In a pluralistic society, whose history and values shall provide the defining symbols that dominate the landscape and make up place, region, and nation?

Meaning and value hold central relevance within research into human-environment relations. However, to create a theoretical basis for such research, one must meet at least two apparently unrelated challenges. First is the root difficulty that discussion of values within social science has always faced: how to legitimate methodologically the taking of a moral position? Second is the difficulty, amounting to an imperative where definition of the research object is a central issue, of avoiding arbitrary and unexamined constructs. Historically, values have been imported into empirical inquiry through reliance on some sort of foundational truth, in other words, a system of assumptions constituting a worldview. The assumptions, rather than the inquiry itself, tend to furnish the research constructs.

To avoid this situation, the theoretical base must specify, and so open to examination and challenge, the researcher's assumptions. The deepest of these assumptions are ontological, that is, related to the nature of being: what it is to conduct research, what constitutes an object, and so on. This is the level on which values are formulated, hence, the implicit claim I make in this essay that a return to ontology is needed. The claim has two parts: first, that unexamined definitions of being have played an important role in the genesis of what are now seen as environmental issues. Second, to the extent that researchers are unable to gain conceptual access to their own definitions and definition process, they stand to further entrench existing, problematic definitions within human-environment relations.

The purpose of this chapter is to develop the above claims using the spatial ontologies of George Herbert Mead and Martin Heidegger in *Being and Time*. The argument presented focuses on the nature of spatiality, bringing together its various aspects and emphasizing its social nature. The argument is intended to contribute to a morally explicit and empirically useful understanding of human-environment relations as they are studied by geographers. An examination of environmental philosophy, or of historical sources of theory in geography, would constitute further contributions. However, these topics are beyond the scope of the present inquiry, in which the philosophical argument itself is the intended contribution.

Experiential Versus Subject-Object Inquiry: Limitations and Challenges

Mead and Heidegger are not directly linked historically, but Mead within pragmatism and Heidegger within phenomenology held in common an emphasis on the processual and experiential. This emphasis developed over time, circa 1870–1940, with a degree of mutual influence between the two philosophical schools. However, the driving impulse behind both lines of inquiry was probably Kant, whose category of the "phenomenal" proposed as early as 1770 was still evoking fertile responses 150 years later.[1] The "phenomenal" indicated objects, events, and so on as they appear to us in experience, in contrast to these entities "as they really are." Mead's, Heidegger's, and indeed Kant's explorations arose from dissatisfaction with the limitations of a Cartesian "scientific method" that excluded experience as an object of inquiry. In regard to the study of human-environment relationships, I argue that this dissatisfaction can only be addressed through an account that experientially defines the three relevant entities: human beings, the environment, and relationship. Mead and Heidegger in effect dissolve and reform the three entities by beginning with "relationship" as it is experienced. The validity of experience is assumed, then used to build up definitions of the phenomena to be related. This process is the opposite of what has often been referred to as the Cartesian mode of inquiry, which assumes a separate subject and object. Subject and object must be joined through the forging of a reductively defined knowledge relation.

The two difficulties I have raised stem from a Cartesian worldview and the positivist methodology that developed from it during and after the Enlightenment. However, restating historical positivist assumptions in a way that highlights their limitations, as I will do below, is not done for the sake of a critique that is hardly needed at this point in social-science history. Cartesianism and positivism have played inestimable roles historically in forwarding inquiry. Marx, who David Harvey has pointed out was in some ways a child of the Enlightenment, was a founder of inequality studies; I shall present Marxian thought here as the strongest challenger to the experiential approach of Mead and the early Heidegger.[2] It would probably be impossible, even if desirable, to discard the analytic framework that Cartesian and Comtean perspectives made possible when they rendered logic operational and connected it methodologically to systematic empirical observation. The object in reexamining parts of the subject-object analytic below is rather to isolate methodological characteristics that fit poorly within certain situations we currently wish to study.

By asserting the validity of everyday observation, Mead and Heidegger both attempt to place back into context rigorously reductive assumptions that made many social phenomena scientifically invisible. Cartesian rationalism and Comtean positivism constructed a set of one-way relationships, first the subject-object relation, then independent-dependent empirical variables. These dichotomies separated knower and known, causer and caused, and constituted both as prima facie entities. The act of constituting rendered context methodologically null and begged the questions of arbitrariness and self-interest in the defining of social entities. In social inquiry historically, the methodology has permitted a figure-ground worldview in which certain human actors (later, abstract social forces) were figures. All else— "environment," nonhuman beings, and nonrelevant human beings— were ground and could not be developed as themes within the inquiry. Further, the subject-object dichotomy, as knowledge relation, posited a "transparent" act of observation. Privileging one set of values over another, though it may have been given the justification of empirical observation, was also a nakedly interpretive act that challenged the assumption of transparent observation. Valuing cannot be formally acknowledged *or regulated* within positivist methodology.

Marx, as a founder of inequality studies, posed ingenious challenges to the limitations of positivist methodology. Marx incorporated an ethical position—egalitarian access to the means of production—by making it the foundational truth toward which inquiry is directed, through the "ethical methodology" of praxis. Within the Marxian epistemology, relations of domination were seen to remain invisible precisely because of a flaw in subject-object knowledge relations: the flaw of "false consciousness." But Marxian ontology, essentially a labor theory of human nature coupled with Hegel's stage model of history, did not concede much either to space and place or to the nonhuman world that has since become part of human-environment inquiry. A clear example of limitation is Marx's treatment of "nature." The earth as a productive support base for "mankind" was assumed as readily by Marx as by John D. Rockefeller. "The celebrated unity of man with nature," Marx remarked in the *German Ideology*, "has always existed in industry [i.e., in human actions upon nature and the corresponding effects of nature's products upon human social life]."[3] Although Marx stressed human beings as a part of nature, this passage and others throughout Marx's work clearly establish nonhuman "nature" as destined to become "one of the organs of human activity," annexed to our own bodily organs.[4]

Marxists have gone beyond Marx, to be sure. Marxian thought has energized the working out of more theoretically complex geographies of uneven development, marginalization, and sustainability; has occasionally even rendered some level of "equity," "autonomy," or "voice" to the nonhuman as well as to persons in their diversity. In *Dialectic of Enlightenment*, for example, Max Horkheimer and Theodor W. Adorno could construct the syllogism that the domination of nature leads to the domination of human beings.[5] As David Harvey has argued, the "totalizing spatial vision" constructed by Enlightenment thinkers and instanced by cadastral systems and depictions of distribution simultaneously recognized cultural diversity and allowed development of a hierarchy of worth or importance among cultures.[6] Harvey has also pointed to spatiality in its aspect of place as one "naturalizing medium" enabling false consciousness to operate.[7] However, the static quality of Marx's moral world continues to limit the objects of inquiry to which the "conflict perspective," in Randall Collins's phrase, can be applied. Marx's methodology provides for manipulating a set of prima facie entities rather than generating objects of inquiry inductively on a criterion of validity. This limitation is representative of the logicopositivist failure to recognize the constructed nature of social objects. In Heidegger's and Mead's claim, construction takes place experientially through the act of relating.

Mead and Heidegger are presented here as radical ontologists who (1) successfully abrogated the Cartesian subject-object dichotomy and (2) used the freedom this gave them to turn away from foundational truth toward a reliance on process. Marxists have raised caveats, justifiably, about the asociality of Heidegger's ontology and about its aestheticism as a questionable basis for social action.[8] I contend that the latter limitation applies chiefly to the later Heidegger. I confine myself to arguing here that Heidegger's method of inquiry, as it is modeled in *Being and Time*, is strong at exactly the points where a logicopositivist methodology is weak. To the extent that Heidegger's ontology is experientially derived, it is able to reunite disparate conceptualizations of space, from place or region to spatiality itself. By contrast, proceeding from a subject-object dichotomy confuses the issue by unjustifiably reifying and compartmentalizing "human beings," "environment," and "relationship." Further, Heidegger's arguments integrate the ethical by grounding a normative dichotomy, authenticity versus inauthenticity, within time and space. I present Mead as supplemental to Heidegger specifically on the point of asociality. Mead's treatment of spatiality is limited almost to a footnote on "environment." Nonetheless, Mead supplies a model of social process that does include the spatial and, importantly, suggests a direction for an ethics of process that avoids reliance on foundational truth.

Relationship and Difference in Pragmatic and Existential Inquiry

Pragmatism and the existential branch of phenomenology are characterized by structural similarity and intentional dissimilarity. These relationships can be illustrated by the respective terms in which the human being as object of inquiry is identified and defined. For phenomenologists as quasi-existential philosophers (e.g., Merleau-Ponty, the Heidegger of *Being and Time*), the human being is "that entity which in its Being has this Being as an issue" (H42/68).[9] For Heidegger, existence precedes essence but does not preclude it (H42/67 and Introduction, Sec. 5–6, H15–27/36–49). By contrast, for the pragmatist, rooted generally in empiricism and specifically in experimental psychology, process itself constitutes the proper object and ultimate reach of inquiry.[10] In pragmatism the human being, insofar as he or she stands apart from other entities, even others that are sentient, is an "individual" or "self." "The self has the characteristic that it is an object to itself."[11] Self, however, arises from and can be fully accounted for by a physical organism.[12] This is a situated rather than transcendent characterization, one that facilitates the pragmatic inquiry centering on the analysis or modeling of experience.

A point of similarity between pragmatism and phenomenology, on the other hand, is the central position that their analyses give to an experiential world whose entities are socially constructed or defined. Both Mead and Heidegger (H59–62/86–90) explicitly make the point that experience itself cannot be empirically investigated from the Cartesian assumption of a subject-object dichotomy.[13] Experience, defined as involvement, interpretation, or meaning, is excluded from inquiry when the subject-object dichotomy delimits the subject as *cogito* and places objectification ontologically prior to other relationships of subject and object. Experience can only be incorporated into the analysis as a knowledge relation that the subject is striving to establish. But the knowledge relation is forever subject to subversion by the unreliable powers of the very subjectivity that strains to perceive the world. The credibility of experience, not its nature, becomes inescapably the goal of inquiry. Further, if experience is not construed as ontologically prior to objectivity, experience must somehow be accounted for as a derivative of objectification. At this point, both experience and intentionality necessarily become metaphysical entities. That is, one must account for them by referring to an outside source, such as God (in Descartes's system) or a nonexperiential quality inhering in the subject (as in Kantian a priori structures of mind). Relevant to inquiries into ethics, experiential and intentional categories such as "domination" that have a great deal to do with personal, social, economic, and political outcomes become very difficult to incorporate into an analysis. To demonstrate experience as formative or constructive, in relation to the built environment or anything else, a nondualistic analysis is required.

Phenomenological and Pragmatic Analysis of Experience

In spite of divergent objectives and sometimes entities of inquiry, phenomenologists and pragmatists have developed positions about the nature of the world, existence, or "reality" that are similar. Neither group engages ontology in a traditional, metaphysical sense. "Reality" is operationally defined as the world of experience. The realness of the world's constituent matter, energy, perceptions, relations, transactions, and so on are taken at face value. Inquiry begins at the point where these constituents of reality are formed into "world" by the human perceiver/actor. For the pragmatist, idealist or subjectivist considerations do not arise. For Heidegger (H205-8/249-52) and some other phenomenologists, the question of the "independent reality" of things and the philosophical pursuit of "proofs of reality" are red herrings. The reasons for this are inherent in the phenomenological definition of being as "Being-in-the-world," which will be discussed below. The experiential definition of reality is not to be confused with the view that the world is subjective, having no other existence than through our perception of it. Nor does the definition admit the Kantian view that our perceptions themselves are prior to or independent of that which is perceived (e.g., H109-10/143-44).

Following from the above, the first point to be made is that for both pragmatists and phenomenologists, what makes reality into "world" is reality's construction into entities by a perceiver who responds to real stimuli. The perceiver's relationship to phenomena determines what shall constitute an entity. For human beings, everything nameable is such a constructed entity, including "being" and "the self." Further, the process of naming itself indicates the social nature of world-construction. Both Mead and Heidegger accept the social nature of "world." Mead gives primary attention to this proposition while for Heidegger the proposition remains latent.

As a second point, the concrete level of experience, as opposed to the abstract or conceptual level, is primary. The "reality par excellence" is everyday life, the phenomenological *lifeworld*.[14] Apprehending everyday life, existing within it, constitutes the "natural attitude." More closely argued versions of this position may be found in the work of both Heidegger and Mead as they respond to the difficulty of making inquiry into experience from a position of Cartesian dualism.

For Mead, "the field of immediate experience is the [individual's] support for getting his lever under the world of reflective thought."[15] All immediate experience arises from the action of the organism in relation to a stimulus. Whether this experience is classified as "sensory" or "mental" does not reflect a difference in kind but depends upon whether the experiencing individual receives the stimulus from an external source or provides the stimulus by addressing him- or herself. The procedure of addressing oneself consists of re-presenting to attention the attitudes or imagery derived from prior experience that was, at some stage, sensory.[16]

Heidegger focuses on the nature of Being as a totality rather than on the nature of the individual. Within the totality of Being, human being is Dasein, the "be-there." Karsten Harries explicates the Heideggerian human being in this way: "To be a self is to experience the things of the world from within the world"; the human being is in the world essentially rather than by happenstance.[17] Human being, says Heidegger, has "proximally and for the most part" the "undifferentiated character of Dasein's everydayness" (H43/69; also H15–17/36–38). Our everyday being has the characteristic of unselfconscious involvement in a life composed of tasks; thus everyday being is based on involvement, or care (*Sorge*) (H57/84). Detachment or "objectivity" is a derivative mode, a conceptual level of being that derives from its experiential one—the level of the "present-at-hand," where phenomena can be abstracted from their totality (e.g., H55ff/81ff; H61–62/88–89). Through a kind of deficient mode of involvement that Heidegger describes as "just tarrying alongside" without "producing, manipulating, and the like," space can become a theme of circumspection and can be calculated and measured (H61–62/88–90).

Relatedness and the Experiential World

In the process of giving primacy to experience over abstraction, both pragmatists and phenomenologists defend the claim that the self and the other are mutually created as entities through a relational process. William James and John Dewey, as well as Mead, developed "the recognition that organism and environment are mutually determinative of each other."[18] For Mead, "the reflective experience, the world, and the things within it exist in the form of situations. These situations are fundamentally characterized by the relation of an organic individual to his environment or world. The world, things, and the individual are what they are because of this relation."[19] Similarly for Heidegger, "it is not the case that man 'is' and then has, by way of an extra, a relationship-of-Being towards the 'world' " (H57/84). Being is a relational totality, "Being-in-the-world" (H52–62/78–90). For Heidegger as for Mead, without "world" there is no possibility of either human being or the human "discovery" of other entities (H73ff/102ff); all three are terms of each other equiprimordially. Within the world are many beings and entities, but their sum total does not equal the world (H72–76/102–7). Ontologically, the world is "worldhood"; for Dasein, that which allows us to encounter otherness as part of our Being-in-the-world. Worldhood comprises kinds of relation such as understanding of purposes (H67ff/96ff); familiarity with relationships (H72ff/102ff); involvement with entities (H83–87/114–21); and solicitude for beings (H117ff/153ff).

Heidegger's analysis of Being-in-the-world as inherently spatial forms the basis of my discussion below of the social role of the built environment. However, because for Heidegger the social nature of relatedness remains implicit, it is necessary first to present Mead's construct of relatedness as social. Mead presents the relationship of individual and social development as a type of discourse.[20] The act is the basic unit of experience, and the act as experience achieves its fundamentally social nature by being communicable. Out of the communicability of the act rises the self, that part of the human organism that is an object to itself.[21] The self is formed as the individual "takes the attitudes of other individuals toward himself within a social environment or context of experience and behavior in which both he and they are involved."[22] It is communication "in the sense of significant symbols," notably language, that enables the individual to objectify him- or herself.[23] Significant symbols make up the script by which one plays the role of the other and is thus enabled to internalize the other's point of view as a "me."

Ultimately, the totality of viewpoints making up the abstract community or social whole is internalized as the "generalized other." This internalization enables one to function as a member of society. Internalization of the generalized other also enables one to conduct an interpretive dialogue within oneself about one's functioning. As a socialized individual, that is, one possessing a fully developed self, "I" see "me" in the sense of both structure and content based on the way that others see me. I have come to possess gestures, attitudes, and goals in common with the rest of my group.[24] As Herbert Blumer emphasizes, this commonality is critical for the meaning-structure that binds together a community or other social group.[25]

How then does "the environment" enter into this meaning-structure? Mead establishes the environment as a socially constructed entity, or object.[26] An object, being anything that can be indicated, can be responded to socially and thus can be internalized as a type of other.[27] The use of the word "object" does not imply that this entity is physical, and as noted above, any classification of objects into corporeal and noncorporeal is a matter of convenience.[28] Paul Tibbetts makes the argument that the difference between physical and social constructs is a matter of context, namely, "social interests, linguistic preferences, and a world-at-hand."[29] In a footnote, Mead sketches the process of incorporating "the physical environment" into social relations:

> the cult, in its primitive form, is merely the social embodiment of the relation between the given social group or community and its physical environment—an organized social means, adopted by the individual members of that group or community, of entering into social relations with that environment, or (in a sense) of carrying on a conversation with it; and in this way that environment becomes part of the total generalized other for each of the individual members of the given social group or community.[30]

But it is Heidegger who explicates the nature of this physical environment so briefly and casually referred to by Mead.

Relatedness as Spatial

If Mead establishes our existence as built up from a process of social relationship, Heidegger establishes the spatial nature of relatedness itself (H104–10/138–44). For Heidegger, the process of making our relationships within the world intelligible to ourselves is a spatial process. As our everyday being has the characteristic of unselfconsciousness involvement in a network of relationships, involvement or care is our very means of spatial apprehension, the means by which we locate everything, including ourselves, within the world. For instance, "close" and "far" are experienced in terms of Dasein's involvement rather than measured distance. In Heidegger's example, the glasses on one's nose, the pavement under one's feet, may be "discovered," or located, as remote compared to the friend sighted twenty feet away, because we are attending to the friend and not to glasses or to pavement (H107/141-42). The closeness of the friend in experiential terms reflects Dasein's spatial attribute of de-severance, which "amounts to making the farness vanish" (H105/130). Dasein is not centered on a point or "here" but is always "yonder," through de-severance, with the beings or entities it is concerned with. Only from these and in terms of its concern with them can Dasein come back to—i.e., establish—a "here" (H108/142-43).

For purposes of deriving an ontology, then, space is not a thing or a relation. Space cannot be a thing, because things exist in space. Space cannot be a relation, because any relational term such as "between" presupposes space. Rather, space is a different kind of entity, arising from the phenomenon of our engagement with the world: Space is the disclosure of location on the basis of significance (H110/145). Through its nature as Being-in-the-world, "Dasein always has space presented as already discovered" (H112/147). Further, the quality of spatiality as an attribute of relational cobeing gives rise to what we may call "institutionalized space," the Heideggerian place or region.

Heidegger argues that spatiality is regionalized, amounting to a process of Being's creating place. Implicitly, this is a social process in which "Dasein" is to be taken as the being-there of more than one individual. Because Dasein apprehends beings and entities in terms of the relatedness of worldhood, it locates them in terms of the region (H102-4/135-38). The region, a less-literal version of geography's "functional region," is the location of a totality of referents necessary to carrying out some task, referents that Heidegger calls the "ready-to-hand," "equipment," or "gear" (secs. 15 and 16).

The point is the items' relatedness: the region must exist for items to be discovered within it (H102-3/136). The region acquires its locales, each item is discovered at a locus, only in relation to all items of a particular referential totality. In everyday life, the region, like space itself, remains unobtrusive and unthematized in "inconspicuous familiarity" (H104/138).[31] Typically, one becomes aware of the region when an item needed for a task cannot be found. One's awareness of region is shown, for instance, by not ordinarily looking for one's reading glasses in the refrigerator. Further, as with the "close" and "far" of Dasein's spatiality, our awareness of region and place begins in experience (H103–4/136–37).

Heidegger's ontology of the spatial might seem a simple elaboration of Mead's were it not for the rather sinister characterization Heidegger gives to the social world and to spatiality itself. Mead's generalized other, as social group, fosters creative individuality and alone brings the self to maturity.[32] It follows on Mead's analysis that place, as a concretization of "the generalized other," can be internalized and become formative for the self. But for Heidegger, to experience the involvement or care of everydayness is to lose oneself. The social interaction that determines our everyday experiential world, including the region or place, is carried out by Dasein in the mode of forgetting itself: "they," *das Man*, the inauthentic self, "the Other that takes away Being" (part 1, chap. 4, esp. sec. 27, H126–30/163–68; H126/164).[33]

In Heidegger's analysis, the spatial structures of the they-self, including the regions determined by care, do not carry the potential for disclosure of authentic Being (H184–91/228–35). Being, "the authentic self," is the self that has found its own nature (H129/167); however, the nature of involvement as spatial is typically that it precedes or bypasses this type of self-encounter. Toward the end of *Being and Time*, Heidegger presents spatiality as the very condition of human self-forgetting (H367–72/418–23). The bringing-close effected by Dasein's spatial nature is also a making-present; as spatial one becomes trapped recursively in a perspectiveless here and now. It is only through the discovery of nothingness, the no-place, that Dasein becomes aware of itself as "the place where beings disclose themselves," that is, where significance or meaning is discovered.[34] This discovery is only to be attained through the act of individual "disclosure," not that of social construction. Authenticity is the condition of our existence as moral beings, says Heidegger, and authenticity implies a resolute individual consciousness freeing itself from the spatial embeddedness of the they-self.

By contrast, Mead notes but does not evaluate the fact that communication by significant symbols need not be a process of consciousness. The ambiguities of Mead's "I" have been pointed out by various commentators. For Mead, of course, the critical issue was not "authenticity" but social cooperation. Mead appeared rarely to feel the need of evaluating social cooperation in a moral sense; he seemed rather to endorse without conflicts the ideals of his nation and society. Thus Mead's moral framework could generally remain at the level of assumption. Mead's focus sets an obvious limit to his analysis of consciousness just as Heidegger's focus sets a limit (about which he is explicit) to his analysis of the social.

Spatial Ontology, Ethics, and the Human-Environment Relationship

The Heidegger of *Being and Time* begins with the concept of Being-in-the-world, or primordial relatedness, and propounds spatiality as a constructive dynamic of Being-in-the-world. In this account, process rather than a priori entity is central to the analysis of social-spatial relations. Mead supplies a format for social process by characterizing the construction of the generalized other as a "conversation." Though Mead hints that place can take part in the conversation, his privileging of the language relation impedes application of his analysis to spatial phenomena. By contrast, as explicated in the preceding section of this chapter, Heidegger privileges spatiality within the act of relating.

Moreover, Heidegger's account of spatiality as the condition of our self-forgetting provides a standpoint for ethical evaluation. The "underdetermined" nature of Heidegger's ethics has often been commented on.[35] I have noted that Heideggerian ethics has been all but dismissed within Marxist theory for its social inadequacy. However, the Heideggerian principle of authenticity is arguably ethical in that it sets up a hierarchy of values for human states of being, within which all moral positions must presumably arise. At the head of this hierarchy is Dasein's mode of authenticity, its resolute, concernful realization of its own nature.[36] Heidegger placed authenticity in relation to ethics when he wrote, "Only from the truth of Being can the essence of the good be thought."[37]

But spatiality, a primordial part of Dasein, does not belong ontologically to the authentic self. And, as David Harvey has observed, "spatial and temporal practices are never neutral in social affairs."[38] Historically, Harvey posits a volatile, ongoing reorganization of space driven by the development of capitalism. Harvey follows Lefebvre in pointing out the tendency toward "homogenization" of space into an abstract commodity, as opposed to a discrete set of fixed and unique locales whose value is in many ways symbolic.[39] Contrasting effects of these two types of space creates a tension. "Pulverization" of homogenized space into parcels that can readily be divided, bought, and sold results in greater freedom of action, for better and worse, expressed partly through chronic jockeying for locational advantage. Place, by contrast, is the container for dynastic, or static, power. Implicitly, place as territory is symbolically or literally "held" by a group or coalition of interests. Their ascendancy is legitimated by a symbol-laden built environment and, I would add, normalized by the large-scale and temporal persistence characteristic of landscapes.

The point of joining Heidegger's analysis to a Marxian one here is that Heidegger accounts for the self-contradictory possibilities that Harvey depicts both in the forms of space and in their ethical implications. Within Heidegger's spatial ontology, all spatial formulations are upheld by their not belonging ontologically to the conscious, self-encountering, or "authentic" mode of being. Without connection to the source of our ethical self-interrogation, our perception of spatiality may create regions applicable to any "task," including tasks of domination and exclusion. Manipulation of location may itself be seen as a regionalizing process, the "disclosure" of proximate connections that create relative advantage.

However, Heidegger's conceptualization of authenticity in *Being and Time* holds an ethical ambiguity that is related to Dasein's spatial nature. As spatial, Dasein repeatedly subverts its potential for authenticity through a tendency to be "sucked into the turbulence of the 'they's inauthenticity."[40] At the same time, given Heidegger's repeated emphasis on "fallenness" as inherent in Dasein, a definitive escape from the inauthentic would be an escape from Dasein's own nature. The later Heidegger attempted to resolve the ambiguity of the authentic through a turn to the foundational, which Ballard notes as beginning in 1930 with "On the Essence of Truth." Heidegger abandoned "the voluntaristic notion of human transformation" earlier implicit in authenticity, moving toward a passive stance wherein a priori entities or "supra-individual forms" shaped the destiny of Being.[41] Coming at the historical moment that it did, Heidegger's earlier, existential analysis was both informed by and ultimately perhaps overwhelmed by the moral ambiguity of human being as spatially situated. As Harvey rightly points out, "place" in Heidegger's thought came to assume a foundational status—for example, in the concept of "the Fourfold"— that admitted aesthetic criteria into the justification of social action and probably facilitated Heidegger's accommodation to Nazism.[42]

It is appropriate to challenge Heidegger's ambiguous definition of the authentic, as has been done both by Marxians and by the phenomenologists whose perspectives uphold Berthold-Bond's critique.[43] However, as Berthold-Bond himself points out, ambiguity contributes to the understanding of authenticity or "resolute consciousness" as a process or struggle. There are implications in the condition of process or struggle for a methodology of the ethical. Pragmatists like Mead as well as post-Marxists have held that foundational truth is an illusory basis for moral choice. If this position is embraced, the nature of moral choice must be seen as permanently emergent and "irredeemably ambivalent."[44] As Dasein, we can never overcome our spatial and temporal situatedness. Nor, within the nature of moral choice itself, is there any actual justification for a definitive act of transcendence. Nonetheless, we are existentially constituted as "the understanding which interprets" (H153/195), and our quest of "overcoming" or transcendence is both our avenue of moral potential and the connection of that avenue to the life of the intellect. Moreover, our intractable ties to "the environment" coupled with our potential for moral self-interrogation provide the ontological basis for an environmental ethic.

Commentators over time have offered a variety of resolutions to the problem of emptiness, formality, or abstraction within Heidegger's concept of authenticity. Because the focus of my analysis has been spatial, I turn to Mead, with his hints of spatiality as part of the social process, in order to supply a dimension of praxis to Heidegger's ontology. Mead's description of the formation of the generalized other can be narrowly interpreted as a process of immediate social consensus within a human group. However, a broader interpretation would take the "conversation" that forms the generalized other as an ongoing quest for empirical and political inclusivity, ranging over both space and time. (As Berthold-Bond suggests, somewhat along the lines of Gadamer, a reflective grounding in the past serves as an indispensable guide to resolute, concernful future choices.) Interpreting the "me" of any individual as a digest of many voices, many "others" (including environmental others), refutes the position of stoic isolation that Heidegger's authentic self tends to fall into. Conversely, the Heideggerian position of resolute self-encounter is needed to overcome the possibility for moral status quo inherent in Mead's construction of social-spatial interaction.

Conclusion

Issues around marginalized human groups, resource distribution, biodiversity and sustainability, and the rights of nonhuman beings highlight the need for a theoretically adequate definition of the human-environment relationship. An ontological construct is needed as a point of departure for inquiry that is morally and intellectually defensible as well as empirically relevant, a basis for action. Mead and the Heidegger of *Being and Time* are indispensable for their making visible spatial-social phenomena not able to be thematized under the strictures of a Cartesian-positivist methodology.

Ultimately, given the arguably divergent objectives of philosophy and of social science, the quest for satisfactory theoretical arguments must reenter the social science stream.[45] An empirically based analysis of intentionality, for example, is needed to explicate the creation of regions as political. However, the spatial ontologies of Heidegger and Mead supply key supports for the social-science arguments from which the "human-environment relationship" must be constructed. The current challenge to foundational truth in general is in part a critique of stasis as a criterion of truth. Any "foundational truth" that is humanly generated necessarily operates from a perspective. As soon as the perspective is "universalized," paradoxically, it becomes frozen in time and space. This feature caused Marx's perspective, for example, to become dated. The drive to universalize also exposed within Marx's perspective a locational bias in the exceptional example of British industrialization. The point is not that Marxian thought is less useful because it is perspectival. Rather, its claim to universality has led to inappropriate applications and weakened the very predictive power that was sought. In Heidegger's and Mead's analysis, motive itself is questionable as an a priori entity; "self-interest" is constructed within time and space. Clearly, for human-environment relations, the problematic remains one of bringing and keeping the perspectival nature of our understanding under scrutiny.

Notes

1. Immanuel Kant, *Kant's Inaugural Dissertation and Early Writings on Space,* trans. Handyside (Chicago: University of Chicago Press, 1929 [Koenigsberg, 1770]). Kant's use of the word "pragmatic" to mean empirically lawful or experimentally derived was claimed as an influence by Charles W. Peirce, founder of American pragmatism. See Peirce, *Collected Papers,* ed. C. Hartshorne, P. Weiss, and A. W. Burkes (Cambridge: Harvard University Press, 1931–1938).

2. Harvey, *The Condition of Postmodernity* (Oxford: Basil Blackwell, 1989).

3. "The German Ideology," in *The Marx-Engels Reader*, 2d ed., ed.

Robert C. Tucker (New York: W. W. Norton, 1978), 170. Also see, in the same source, the *Economic and Philosophical Manuscripts of 1944*, 87–88, 90, 115ff, *Grundrisse*, 410, and *Capital* 1: 328, 345ff.
   4. *Capital* 1: 345 in Tucker edition.
   5. Max Horkheimer and Theodor W. Adorno, *Dialectic of Enlightenment*, trans. John Cumming (New York: Herder and Herder, 1972), 13.
   6. Harvey, *Condition of Postmodernity*, 250–52.
   7. E.g., David Harvey, "Monument and Myth," *Annals of the Association of American Geographers* 69 (1979): 312–81.
   8. Theodor W. Adorno, *The Jargon of Authenticity*, trans. Knut Tarnowski and Frederic Will (Evanston, Ill.: Northwestern University Press, 1973); Georg Luckács, "Existentialism or Marxism?" and Herbert Marcuse, "Sartre, Historical Materialism, and Philosophy," in *Existentialism versus Marxism*, ed. George Novack (New York: Dell, 1966); Harvey, *Condition of Postmodernity*, 207ff, 304.
   9. Martin Heidegger, *Being and Time*, trans. from *Sein und Zeit* (1926) by J. Macquarrie and E. Robinson (New York: Harper and Row, 1962). Because of frequent references to *Being and Time*, page numbers here and following are given in text. Numbers prefixed with H refer to the 1926 German edition, numbers after the slash mark refer to the translation.
   10. George Herbert Mead, *The Philosophy of the Act*, ed. Charles W. Morris, J. M. Brewster, A. M. Dunham, and D. L. Miller (Chicago, Ill: University of Chicago Press, 1938), esp. xvi.
   11. *The Social Psychology of George Herbert Mead*, ed. Anselm Strauss (Phoenix Books, Chicago, Ill: University of Chicago Press, 1956), 213.
   12. E.g., Mead, *Philosophy of the Act*.
   13. Mead, "Consciousness, Mind, the Self, and Scientific Objects," 176–96 in David L. Miller, ed., *The Individual and the Social Self* (Chicago: University of Chicago Press), 179–84; also see William Barrett, Introduction to "Phenomenology and Existentialism," in *Philosophy in the Twentieth Century: An Anthology*, vol. 3, ed. W. Barrett and H. Aiken (New York, NY: Random House, 1962), 135–36.
   14. Peter Berger and Thomas Luckmann, *The Social Construction of Reality* (New York: Doubleday/Anchor Press, 1967), 21ff.
   15. Mead, "Consciousness," 183.
   16. Mead, "Consciousness," 177.
   17. Karsten Harries, "Fundamental Ontology and the Search for Man's Place," in *Heidegger and Modern Philosophy* (New Haven: Yale University Press, 1978), 68.
   18. Morris et al., eds. *Philosophy of the Act*, vii; also see Miller, ed., *Individual and the Social Self*, 19.
   19. Mead, *Philosophy of the Act*, 215.
   20. Mead, *Social Psychology*, 128–35, 212–29.
   21. Mead, *Social Psychology*, 213.

22. Mead, *Social Psychology*, 215.
23. Mead, *Social Psychology*, 216.
24. Mead, *Social Psychology*, 229–232.
25. E.g., Herbert Blumer, "Symbolic Interactionism," in *Three Sociological Traditions: Selected Readings*, ed. Richard Collins (Englewood Cliffs, N.J.: Prentice-Hall. 1985).
26. E.g., Mead, *Philosophy of the Act*, 271ff.
27. Mead, *Social Psychology*, 231, note 7; also see quotation below.
28. Blumer, "Symbolic Interactionism," 289; Mead, *Philosophy of the Act*, 271–273.
29. Paul Tibbetts, "Threading-the-Needle: The Case For and Against Common-Sense Realism," *Human Studies* 13 (1990): 309–22.
30. Mead, *Social Psychology*, 231, n. 7.
31. Cf. Sec. 16 on the inconspicuousness of the ready-to-hand.
32. Mead, *Social Psychology*, 262–63.
33. On Heidegger's equating everyday involvement with inauthenticity, see Daniel Berthold-Bond, "A Kierkegaardian Critique of Heidegger's Concept of Inauthenticity," *Man and World* 24 (1991): 119–142.
34. Harries, "Fundamental Ontology," 68; cf. Derrida's various discussions of "absence."
35. E.g., B. W. Ballard, "Marxist Challenges to Heidegger on Alienation and Authenticity," *Man and World* 23 (1990), 125ff.
36. For sympathetic and less sympathetic treatments of Heideggerian authenticity as ethical, see Michael Zimmerman, "Karel Kosik's Heideggerian Marxism," *The Philosophical Forum* 15 (1984), esp. c. 220, and Daniel Berthold-Bond, "A Kierkegaardian Critique of Heidegger's Concept of Authenticity," *Man and World* 24 (1991): esp. 125ff.
37. Heidegger, "Letter on Humanism," in *Basic Writings*, ed. David Krell (New York: Harper and Row, 1977), 230.
38. Harvey, *Condition of Postmodernity*, 239.
39. Harvey, *Condition of Postmodernity*, 254ff. Also see Barney Warf's formulation, the "space of places" versus the "space of flows." Warf, "Structuration Theory and Electronic Communications," in *Marginalized Places and Populations: A Structurationist Agenda*, ed. David Wilson and James O. Huff (Westport, CN, and London: Praeger, 1994), 54.
40. H179 (one of many examples of this position), quoted in Berthold-Bond, "A Kierkegaardian Critique," 125.
41. Ballard, "Marxist Challenges," 133. Ballard notes the later Heidegger's partial resemblance to Marx in this respect and ascribes it to "their common debt to Hegel."
42. Heidegger, "Building Dwelling Thinking," in *Martin Heidegger: Basic Writings*, ed. David F. Krell (New York: Harper and Row, 1977); Harvey, *Condition of Postmodernity*, 351; also 295–96 and 303–4 on the sinister ideological properties of space generally.

43. Harvey, *Condition of Postmodernity*, 125–30.
44. Zygmunt Bauman, "Morality without Ethics," *Theory, Culture, and Society* 11 (1994): 1–34; 31 for quoted phrase.
45. Agnes Heller, "From Hermeneutics in Social Science toward a Hermeneutics of Social Science," *Theory and Society* 18 (1989): 294.

Critical Reflections on Biocentric Environmental Ethics: Is It an Alternative to Anthropocentrism?

Roger King

Environmental philosophy engages in critical reflection on the moral rules, conventions, and shared meanings that inform human inquiry and our relationships to nature. As a relatively new branch of moral philosophy, environmental ethics arises out of increasing unease with the destructive environmental consequences of human economic, technological, and cultural practices. Many philosophers are unpersuaded that reliance on traditional utilitarian and rights-based deontological moral theories can adequately respond to the problems, both philosophical and practical, that need to be addressed. There is, therefore, a growing absence of consensus about the underlying moral framework that should define and structure the practices of knowing, valuing, and using nonhuman nature.

Our knowledge of nature is produced as a social enterprise. Behind each inquiry into nature is a communication community whose shared conventions help to structure the conceptual and practical framework within which a person is enabled to investigate nature and share the results of that investigation with others.[1] Shared conventions and meanings also inform the other practices—productive, recreational, aesthetic, and spiritual—through which individuals enter into relationships with the natural world. Among the shared conventions and background beliefs of a particular communication community are certain moral frameworks that structure the moral dimension of these social practices. What are the responsibilities of the investigator or the producer and to whom are these responsibilities owed? What may be investigated, transformed, or created and with what techniques? And what are defensible goals and intelligible motivations for relating to nature within particular practices? Human social practices inevitably presuppose some set of shared conventions about the moral parameters within which knowledge, production, recreation, or contemplation should be undertaken and practiced.

Within the philosophical context of environmental ethics, we are faced with the central task of reflecting critically on the conceptual resources deployed by our society and communication communities. While human relations to nature are not determined exclusively by the ideas and theories that give meaning and significance to our actions, these ideas do offer an important point of entry into what Langdon Winner, following Wittgenstein, refers to as "forms of life."[2] Thus the initial moral philosophical question is, What should our shared conventions and background beliefs be? How should we write the rules for the language games governing our discourses about nature and our relationships to it?

The traditional anthropocentric set of moral conventions privileges human preferences, interests, and needs over those of nonhumans. When costs and benefits are assessed, anthropocentric practices weigh the harm to nonhumans or the disruptions of ecosystems only insofar as these constitute a cost or benefit to human beings. While some environmental philosophers prefer to refine the anthropocentric tradition by sharpening our understanding of the costs to humans of careless destruction of the environment, others focus on articulating a moral perspective that might counteract dominant cultural perceptions of nature.

In this chapter, I wish to delineate two different tendencies within environmental ethics. Running throughout the history of Western culture is a stream of anthropocentric thought that emphasizes human distinctness from nature. Aristotle expressed it when he claimed that the human function and highest good was to be found only in the development of reason, that part of the human soul that is not shared with plants or animals. Descartes reaffirms the tendency in his identification of the individual's essential self with an unextended mental substance, metaphysically alien to the rest of the natural world. Kant and contemporary rationalists confirm the tendency when they identify human free will and autonomy as the cornerstone of moral standing and dignity.

A second, biocentric, tendency exists in contemporary environmental ethics. Deep ecologists, for example, have repudiated the shallow or reformist ecology of human self-interest that seeks only an amelioration of the human impact on nature. For J. Baird Callicott and many others, it has seemed evident that a conceptual and attitudinal switch from anthropocentric to biocentric moral perspectives is essential to an adequate restructuring of moral background beliefs and conventions. Tom Regan has argued that a necessary condition of any environmental ethic is that it defend the existence of intrinsic value in nonhuman nature.[3] This amounts, minimally, to the claim that nonhuman nature has value that is not merely instrumental; nature should be valued for its own sake. This definition of environmental ethics as a biocentric ethics is echoed by Eric Katz. Katz expresses the biocentric orientation to nature as one that privileges untouched wilderness over humanly manipulated landscapes:

> In the context of environmental philosophy, domination is the anthropocentric alteration of natural processes. The entities and systems that comprise nature are not permitted to be free, to pursue their own independent and unplanned course of development. . . . Wherever it exists, in nature or in human culture, the process of domination attacks the pre-eminent value of self-realization.[4]

In this passage, Katz sets up an opposition between the works of human beings and a free, wild nature that has not been altered by human practices. Indeed, any alteration of nature by human beings appears here as a form of domination, and hence as something to be resisted.

The biocentric perspective challenges the anthropocentric assumption that membership in the moral community is restricted to human beings or other (usually fictitious) rational beings.[5] In other words, humans are not the only ones who possess intrinsic value and deserve the respect of moral subjects. This decentering of the moral community away from the human species requires that we learn to integrate nonhuman needs and interests into human moral deliberation.

For those who seek a break with the anthropocentric framework for extending moral consideration to the natural world, the biocentric perspective appears to be a promising alternative. To some degree, however, the biocentric alternative reproduces the dualistic either-or that some philosophers have argued is implicated in human conceptual neglect of the environment. Its negation of anthropocentrism appears to shift moral priority to the well-being of nonhuman beings and communities while obscuring the moral status of humans and their presence in nature.

I shall articulate two different problems with the biocentric alternative to anthropocentrism in order to help motivate a search for a different conceptual scheme. The first is both social and ecological. The biocentric shift away from the works of human beings appears to avoid rather than to deal directly with the fact that humans inhabit and relate to a variety of domesticated as well as wild landscapes. If Katz is right to see domination in the alteration and management of nonhuman nature, then the biocentric perspective would appear to have little to say about how we should structure our relationship to those domesticated landscapes, including cities, rural areas, and agriculture, that have already been altered by human practices.

The domesticated landscape poses a significant challenge to the potential of biocentric environmental ethics. The need to develop urban areas that are in sound ecological relation to their environments, as well as the need to develop sustainable forms of agriculture, forestry, and fisheries highlight the importance of moral reflection on human practices in the domesticated landscape as well as in the wilderness.[6] Agriculture, for example, is by its very nature a human intrusion into wild biotic communities. Some writers, such as Paul Shepard, have even maintained that it is the invention of agriculture that begins the process of human simplification of the environment that culminates in our present environmental crisis.[7] The biocentric emphasis on respect for the intrinsic value of untouched natural systems and processes would appear to minimize the impetus toward ethical reflection on agriculture. If biocentrists are right to charge that nonhuman nature tends to fade out in the course of anthropocentric discourses, one might argue that human beings and their artifacts are equally insubstantial in biocentric theory.

It might be objected here that biocentrists have overdrawn the existence of a gap between wilderness and the humanized landscape.[8] No wilderness, in the sense of a landscape untouched by humans, exists anymore and some landscapes once thought to be wilderness were in fact artifacts of human intervention.[9] The gap between domesticated and wild appears to be one of degree rather than of essence. Nonetheless, the ethical question remains: how should we conceptualize this continuum and where ought we locate ourselves and our impacts on it? It is in order to pursue this question that this essay articulates the two sets of difficulties with the biocentric perspective.

The second set of difficulties with biocentrism is epistemological. Biocentrism, to borrow from Donna Haraway, is a "way of being nowhere while claiming to be everywhere equally."[10] Haraway's remark is intended as a description of relativism. Relativists are in the paradoxical position of being able to value all normative positions equally only by erasing their own position. Were the relativists to take their own moral or epistemological beliefs seriously, these beliefs would begin to structure their evaluations of competing beliefs and the relativist strategy would evaporate. The biocentric theorist may be in the same paradoxical position. This strategy in environmental ethics self-consciously rejects the privileged position accorded to human beings by anthropocentrism and, at the same time, locates equal intrinsic value in all living things or in ecological communities. But like the relativists' version of egalitarianism, biocentric egalitarianism appears to depend upon the erasure of the theorist's own standpoint. This erasure obscures conflicts between human beings as well as the myriad conflicts that constitute natural existence. And just as the relativist will eventually reveal his or her commitments to the careful observer, despite all efforts to avoid privileging any position, so will the biocentrist eventually reveal the point where biocentric egalitarianism breaks down in favor of a privileged individual, species, or ecosystem.

The elusiveness of the theoretical standpoint of biocentrism mirrors the absence of the human figure in the biocentric landscape. Thus, both of the difficulties I have sketched for biocentrism are related. On the one hand, how should we close the ethical gap between the domesticated and the wild landscape, the landscape altered by the human presence and the one left unmodified by human beings? On the other hand, how should we identify the theoretical standpoint from which we might answer the question?

In what follows I shall look more closely at both anthropocentric and biocentric constructions of our moral background beliefs to clarify the difficulties with each approach. I shall then consider an ecofeminist alternative that seeks to ground both the subject and the object of environmental ethical reflection in narrative and ongoing relationships.[11] This alternative calls into question the entire framework of the anthropocentric-biocentric debate.

Why Not Anthropocentrism?

The biocentric orientation to nature is rejected by anthropocentric theorists who, like William Baxter, reduce the value of nonhuman beings to their instrumental, demand value for humans. Baxter writes, "damage to penguins, or sugar pines, or geological marvels is, without more, simply irrelevant. One must go further . . . and say: Penguins are important because people enjoy seeing them walk on rocks. . . . I reject the proposition that we *ought* to respect the "balance of nature" or to "preserve the environment" unless the reason for doing so, express or implied, is the benefit of man."[12]

Since this anthropocentric orientation arguably has had, and continues to have, widespread support as a background assumption of many Western practices affecting nature, we might ask what exactly is wrong with reliance on traditional anthropocentric ethics. Clearly, the anthropocentrist does not face the problem of reconciling the domesticated landscape with the wild the way the biocentrist does. Both landscapes derive their value from their usefulness for satisfying human preferences and interests. Wilderness may be protected for anthropocentric reasons if enough value is placed on its existence. Similarly, forests and farmland may be handled sustainably if such practices satisfy human demands more than other practices would. But for the biocentrist, this contingency is morally problematic.

Without denying that anthropocentrism can become much more environmentally informed and sophisticated, there are still several reasons for suspicion that motivate biocentric ethics. First, it might be argued that without a radical shift in attitudes and beliefs about the value of nonhuman nature, narrowly conceived and short-term human interests will continue to prevail at the expense of the environment. Our sense of difference from and superiority to nonhuman nature is so fundamental to our cultural outlook, it might be argued, that nothing short of a shift to a biocentric standpoint will be sufficient to protect even human needs and interests. From this standpoint, it is essential to develop and adopt a biocentric environmental ethic even in order to promote human rights or preference satisfaction.

A second argument is that anthropocentrism simply fails to articulate the experience of many human beings. Just as many men and women care about their fellow human beings, respect human rights, and hope to minimize human suffering, so too they care about what happens to domesticated and wild animals, natural ecosystems, and the planet as a whole. And while some may see their moral concern as entirely derivative from their concern for human beings, in the Kantian fashion, many others value nonhuman nature for its own sake and not for the sake of other human beings. The phenomenological reality of this experience and the potential for expanding it justifies efforts to articulate an environmental ethic that does not ultimately reduce value to some derivative of human rights and preferences.

A third argument in favor of abandoning anthropocentric ethics is a practical one. If the goal of public policy is simply the satisfaction of human interests, then the resolution of policy conflicts reduces to a balancing of human rights and utilities. In such circumstances, environmental policy may tend to provide less protection both to nature and to human beings than might have been achieved by a biocentric ethic. Eric Katz and Lauren Oechsli have suggested that if the intrinsic value of nonhumans is granted by the parties in policy conflicts, then resolution of the conflicts will also take into account the consequences for nature.[13] Christopher Stone has defended the idea of granting natural entities legal standing on the grounds that unless the natural entity is represented in court proceedings, it is unlikely to benefit directly from damages awarded or reparations imposed by the courts.[14] In sum, the scepticism about anthropocentrism lies in the concern that the definition of costs and benefits will inevitably skew moral deliberations in a self-serving, anthropocentric direction unless we can develop a satisfactory biocentric environmental ethics.

There are, therefore, *prima facie* reasons for taking seriously the biocentric project of locating a form of intrinsic moral value in nonhuman nature, independent of the interests, preferences, and rights of human beings. Nonetheless, it might be argued that a more sophisticated anthropocentrism is conceivable that would entail a less radical disruption of the underlying conventions structuring our relations with other people and with nonhuman nature. I shall briefly examine a position developed by Bryan Norton in order to suggest some of the problems faced even by a more sophisticated form of anthropocentric theorizing.

In *Why Preserve Natural Variety*, Bryan Norton distinguished between a strong and weak version of anthropocentrism and defended the latter.[15] The strong version of anthropocentrism restricts value to the "demand values" of humans. Thus, other species, biodiversity, habitat protection, clean air, or wilderness only have value to the extent that humans actually value them, or, as Norton put it, have a "felt preference" for those things.

Weak anthropocentrism builds on the notion that some felt preferences are irrational or self-contradictory. Value is said to hinge not on felt preferences, but on "considered felt preferences," that is, those felt preferences that survive a process of criticism, reflection, and analysis. Thus, Norton argued, wild nature should be preserved on the grounds that wild things have "transformative value." Possessing transformative value, wild nature provides occasion for human beings to reconsider and evaluate their felt preferences, thereby transforming them into rationally defensible preferences. While Norton applied this analysis to the protection of wild diversity, it offers a strategy that is applicable to the protection of domesticated landscapes as well.[16]

Many people worry that contemporary urban dwellers, particularly children, do not know where their food, water, and material goods come from, and hence do not value farmland, forests, or sustainable production methods. Under these circumstances, farms might be seen to have transformative value as locations for learning how food is produced. Analogously, urban community gardens have transformative value as places for engaging with and valuing natural processes in urban settings, while enhancing social values such as self-sufficiency and a sense of community as well.

The anthropocentric argument for protecting wild species and places because of their transformative value is attractive. And the extension of this approach to the domesticated landscape provides one basis for a defense of open spaces in cities, preservation of historic districts, and exposure of urban children to rural and farming experiences. Despite the attractiveness of the concept of transformative value, I wish to suggest that its abstractness makes the concept epistemologically problematic.

First, transformative value does not reside in objects or places independent of those who experience them. The transformation, and hence the value of what initiates the transformation, depends as much on what subjects bring to their experience as it does on what is experienced. Cultural or class differences, for example, might be expected to lead people to perceive nature differently. If so, the claim that particular places or things have transformative value would have no clear meaning until contextualized for a particular group of people at a particular place and time. Marti Kheel exemplifies this difficulty when she proposes that those who see nothing wrong with eating meat should visit a slaughterhouse.[17] She appears to think that this experience will provide the emotional jolt necessary to unsettle those who use abstract moral reasoning to justify the consumption of meat. But some people visit slaughterhouses and are unmoved. Some people will be transformed by their visit to the slaughterhouse and others will not since the slaughterhouse experience does not have transformative value for everyone. The same point can be made about attitudes toward wilderness. Wild places do not in themselves bring about the experiential transformations Norton hoped for. After all, some people hate the experience (paradigmatically, early European colonists thought the wilderness was the domain of Satan).

A second objection derives from the multiplicity of things that have transformative value. Norton used the concept to argue that wild nature has value independent of merely felt preferences, because the experience of wild nature can transform careless and irresponsible attitudes to nature into insightful and caring action. Transformative values can transform an irrational worldview based on consumption and unlimited economic growth into a rational worldview of environmental responsibility. However, it is also true that the experience of city life can transform the outlook of a person who comes from the country. And contact with American commercialism rarely leaves Third World or indigenous people untransformed. All these experiences would have a transformative value, yet surely there are some transformations we should not encourage. Thus, it remains to be explained why transformations in favor of wilderness protection or farmland preservation are to be valued as stages on the way to a more rational worldview, while transformations that industrialize and Westernize indigenous people are unacceptable. Ultimately, Norton appears to have presupposed criteria of judgment other than the transformative effect of what he values. And for the biocentrist, one plausible conclusion is that those transformative experiences that favor wilderness protection are valuable because wild things after all do have intrinsic value.[18]

The conception of nature as having transformative value is important and a useful tool for thinking about preservation of the domesticated as well as the wild landscape. However, as I have noted, questions remain regarding the standpoint from which transformative value is to be discovered and experienced. There are, after all, radical incommensurabilities between people's perceptions of their environments that compromise the practical efficacy of appealing to transformative value as an anthropocentric defense of wild nature.

This discussion of Norton's position in *Why Preserve Natural Variety* indicates how an anthropocentrically based moral approach to valuing nonhuman nature might support protection of both wild and domesticated landscapes. At the same time, it indicates some of the difficulties that remain. Anthropocentric positions have the challenging task of articulating the moral value of the domesticated landscape and wilderness, while responding to the increasing cultural sense that nature should be dealt with responsibly and intelligently for its own sake. Such positions must also respond to the biocentric challenge that human beings think rather too highly of themselves and that centering an ethic entirely on the transformative experience of an abstract and "unmarked" individual may underdetermine what kind of environmental ethic will be articulated.

The Biocentric Alternative

The biocentric alternative to anthropocentric ethics proposes a more radical reorganization of the underlying conventions constituting human practices affecting nature. If these conventions were to change, we would expect to see important re-evaluations of human conduct, both toward nonhuman nature and toward human beings themselves. Changes in intersubjectivity, held meanings and in the practices they help to define might echo those transformations described by Carolyn Merchant in *The Death of Nature*.[19] Merchant argues that prohibitions or inhibitions against mining grew out of shared understandings of the earth as female. This shared understanding made it possible to interpret mining as a form of violence against the earth. Replacing organic with mechanistic interpretations of nature was crucial, she argued, for the development of scientific and industrial practices that define the modern era.[20]

Biocentrism may be seen to divide into two main strands. One strand is individualist; the other, holistic. I shall argue that both forms of biocentrism fail to bridge the gap between domesticated and wild landscapes and thereby fail to situate the human species in an appropriate place in nature. This failure to situate human beings in nature reflects, as I have suggested, the abstractness of the theoretical standpoint from which biocentrism is advanced.

Paul Taylor's version of biocentric individualism is one of the most detailed. In *Respect for Nature*, Taylor argues that human beings share equal intrinsic value with nonhuman living things.[21] To possess intrinsic value, a being must be what Taylor calls "a teleological center of life." That is, it must have a good of its own that it pursues as best it can. Human beings count as paradigm examples of teleological centers of life. However, Taylor argues that all living beings possess this same quality. A "living thing is conceived as a unified system of organized activity, the constant tendency of which is to preserve its existence by protecting and promoting its well-being."[22] Taylor goes on to say, "No living thing will be inherently superior or inferior to any other, since the biocentric outlook entails species-impartiality. All are then judged to be equally deserving of moral concern and consideration."[23] If all living beings have equal value, then we have a moral obligation to respect that equal value, no matter where it resides. It follows that we should not harm a nonhuman being by interfering with its pursuit of its own good.

Taylor recognizes that biocentrism cannot stop there. Living beings often have conflicting interests. The notion of equal intrinsic value, like Deep Ecology's principle of biospheric egalitarianism, cannot be applied in practice; it is an egalitarianism "in principle."[24]

According to Taylor, conflicts between equally valuable living beings must be resolved with reference to the importance of the interests at stake. An interest is more or less basic, for Taylor, depending on the "comparative importance" of the interest to the organism. In the case of human beings, "basic interests are what rational and factually enlightened people would value as an essential part of their very existence as *persons*."[25] Among the basic interests of persons are subsistence, security, autonomy, and liberty. Despite his biocentric orientation, Taylor believes that both basic and nonbasic human interests frequently take priority over the basic interests of nonhumans. For example, the following nonbasic interests are said to be compatible with respect for nature: building an art museum where natural habitat must be destroyed; constructing an airport, railroad, harbor, or highway involving serious disturbance of a natural ecosystem; replacing a natural forest with a timber plantation; and damming a free-flowing river for a hydroelectric plant. As Taylor puts it, "Whether people who have a true respect for nature would give up the activities involved in these situations depends on the value they place on the various interests being furthered."[26] While further conditions and qualifications might be added, Taylor's argument licenses extensive interference with the natural world, based on the self-conception of their interests by the rational and enlightened person, despite the attribution of intrinsic value to all living individuals.

Does a biocentric individualism bridge the gap between domesticated and wild landscapes? In principle, Taylor's position identifies intrinsic value in all living individuals, both wild and domesticated. Thus, domesticated animals and plants would share equal moral value with their wild counterparts and with human beings. On the other hand, Taylor has acknowledged the conflicts between human and nonhuman interests and stepped back from the full implications of biocentric egalitarianism. But rather than offering a moral framework for thinking coherently about our moral relations to both domesticated and wild landscapes, Taylor deflects the problem of moral choice onto the shoulders of that abstract, "virtual" individual, the "rational and factually enlightened person." Sadly, from Taylor's examples, it is not clear how the "rational and factually enlightened person" differs from the more mundane anthropocentrist on the street.

The difficulty here is not just that Taylor has extended intrinsic value equally to all living beings with one hand and then taken it back with the other—though I would agree with William French that this is an unsatisfactory way to construct a moral theory.[27] Taylor is, after all, quite right to realize that human beings must disrupt nonhuman lives and communities and that doing so with respect for nature is preferable to doing so with regard only to human preferences. The difficulty in Taylor's position is its lack of contextual definition. Who is the "rational and factually enlightened person"? Who decides the terms on which conflicts of interest will be resolved? This disembodied standpoint is a view from nowhere that, in principle, is everywhere equally, but in practice is located somewhere never fully defined. In other words, Taylor's biocentric judgment that all living beings have equal moral value relies upon a bracketing of the conflicts between the interests of humans and nonhumans. But when these are returned to the picture in practice, one is forced to acknowledge that the conflicts are particular to specific individuals and relationships. Taylor does not, however, "embody" or "situate" these conflicts, but leaves them to the disembodied "rational and factually enlightened person" to resolve. This person is not defined, hence, no standpoint is actually offered for resolving conflicting interests.

The holistic version of biocentrism constructs the moral framework for human relations to nature in ways that are relevantly different from those of individualist theorists like Taylor. J. Baird Callicott has interpreted and deployed Leopold's land ethic in defense of a holistic, biocentric ethic.[28] In a section of *A Sand County Almanac* entitled "The Land Ethic," Aldo Leopold outlines a holistic ethical principle that has proven remarkably influential. The land ethic principle states that "a thing is right when it tends to preserve the integrity, stability, and beauty of the biotic community. It is wrong when it tends otherwise."[29] Callicott interprets this principle as the holistic claim that moral value lies in the health of the community, not in its individual members. The value of individuals is, then, derivative of their role in contributing to a healthy biotic community.

In "Animal Liberation: A Triangular Affair," Leopold's views are used to critique the animal liberation movement and the value of the domesticated sphere generally. Callicott writes,

> Leopold's prescription for the realization and implementation of the land ethic—the appraisal of things unnatural, tame, and confined in terms of things natural, wild, and free—does not stop . . . with a reappraisal of nonhuman domestic animals in terms of their wild . . . counterparts; the human ones should be similarly reappraised. This means, among other things, the reappraisal of the comparatively recent values and concerns of "civilized" Homo sapiens in terms of those of our "savage" ancestors.
> . . . The land ethic requires a shrinkage, if at all possible, of the domestic sphere; it rejoices in a recrudescence of wilderness and a renaissance of tribal cultural experience.[30]

In a later essay, Callicott achieves a degree of reconciliation with the animal liberation movement by accepting Mary Midgley's argument that humans do have moral obligations to domesticated animals on the grounds that these animals have become members of a mixed human-animal community.[31] The critique of factory-farming and animal abuse can be based, then, on the notion that humans have "broken trust" with these animals and broken "a kind of evolved and unspoken social contract between man and beast."[32] While this reconciliation mitigates Callicott's earlier hostility toward domesticated animal "machines," it reinstates the divide between the domesticated and the wild. In domesticated communities, a kind of individualist ethic is acceptable that recognizes obligations to individual animals. However, the holistic land ethic should apply to wild biotic communities.

Callicott's solution to the discontinuity of the wild and the domesticated is partial at best. Although the two dimensions are identified, they are not integrated. Domesticated animals are protected by an animal rights ethic consistent with human moral traditions, but other elements of the domesticated landscape are not addressed. Indeed, the holistic ethic appears to require that we evaluate human use of the land in terms of its effects on the integrity of wild communities. Since human action always destroys the wild, untouched character of a community, the biocentric holist is obliged to see humans as destructive interlopers rather than as beings with a legitimate place in nature. The very process of domestication has removed humans, animals, and places from the wilderness and thus runs afoul of an ethic that requires us to protect the integrity and stability of wild ecosystems.

Both Paul Thompson and Gary Comstock have emphasized the irrelevance of a biocentric holism to questions arising in the practice of agriculture.³³ Defense of the integrity, stability, and beauty of the biotic community necessarily entails elimination or maximum restriction of agriculture as long as the focus is on the protection of wild nature. Once again the human figure is forced to beat a retreat from the domesticated as well as the wild landscape. And as before, we may ask, what is the standpoint from which this moral perspective is developed? Who is the theorist who has the luxury, we might say, to argue that people should remove themselves from agriculture and from the transformation of wild nature? According to Ramachandra Guha, this approach to environmental ethics reflects a peculiarly American relation to wilderness, reflecting both the unique history of European occupation of North America, the size of the country and its population, the richness of its resources, and the strength of its economy.³⁴ For most other countries, a holistic biocentrism is either anachronistic because no wilderness remains, or an argument for expropriation that excludes local people from environments in which they have always lived.

What kind of environmental ethics, then, holds out hope for a coherent assessment of our moral relations to both wild nature, agriculture, and the urban landscape? Biocentric holism appears to extend maximal protection to wild nature while at the same time devaluing human culture and individual moral worth. From this perspective, human cities, agriculture, and civilization generally can only appear as a threat to the intrinsic value of ecosystemic wholes, rather than as integral parts of nature themselves.

The upshot of biocentric variations of environmental ethics is that humans do not belong in nature and our presence can only be accommodated by unjustifiable loss to others. Recognition of human needs and interests is generally reintroduced by way of exceptions, qualifications, or *post hoc* limits placed on the initial biocentric ethic. While biocentrism plays an important role in compelling us to re-think the contemporary compulsion for economic development, wasteful production, and ignorant expansion, it provides little positive guidance on how humans might live harmoniously and legitimately within a nature that contains both human and nonhuman aspects.

Value in Narrative and Relationship

I have argued that neither anthropocentrism nor biocentrism succeed in constructing an environmental ethic that identifies and integrates the human place within the domesticated and wild landscapes. Both approaches have focused on privileging one side or the other of the human-nature dualism. In the process, however, either nature or the human, as autonomous, self-directed, and integral domains, is erased. In this erasure, the standpoint of the theorist or moral judge is also left unmarked. It is not expected that gender, class, race, nationality, ethnic background, or historical time are relevant to the validation of either anthropocentrism or biocentrism. The theories become, therefore, views from nowhere, the products of unmarked, disembodied inquirers.

The purpose of the foregoing critique of anthropocentrism is to motivate and justify the search for an alternative ethical approach that is both contextually situated and capable of encompassing the moral perplexities facing us in both domesticated and wild landscapes. Such an ethic will take seriously the concept of place, both as an epistemological category grounding the practice of moral inquiry and as a geographical category situating the precise details and conditions of moral and social action.

One dimension of such an ethical strategy can be found in Karen Warren's notion of a narrative relational ethic.[35] Warren's conception of a relational ethic has two important features. First, it suggests that the locus of value is the relationship rather than the beings who stand in relation. Rather than theorize about the superiority or inferiority of humans and nonhumans, it might be argued, we will do better to reflect on the relationships themselves. Are they oppressive or mutually enriching? Are they sustainable or self-destructing? Do they reflect care and respect or arrogance and domination? Are the relationships worth protecting or should they be terminated? Questions such as these may be asked about relationships both to wilderness and to domesticated urban, rural, agricultural, and resource landscapes as well.

Second, Warren's approach emphasizes the importance of first-person narrative as a vehicle for situating and contextualizing judgments of value as well as knowledge claims.[36] Warren's use of narrative reflects a feminist and contextualist suspicion of the abstract view from nowhere. As Haraway argues, the unmarked knower or judge is not accountable for their theoretical constructions or for the consequences those constructions may have for others.[37] The view from nowhere hides its actual situatedness rather than transcending contextual limitations. Thus, an ethic that makes use of narrative as the vehicle for teasing out moral problems and solutions appears attractive since it can be located in the standpoint of the storyteller.

Warren's conception of environmental ethics is contextualist both in that it sees moral commitments as arising from specific situations and in that it rejects the more abstract search for a single, universal position from which to evaluate or prioritize all our moral obligations, narratives, and relationships. The consequence is that conflicts among humans and between humans and nonhumans must be dealt with in a more *ad hoc* manner as they arise in lived experience. The extent to which mutually satisfying resolutions of conflicts can be achieved is itself a contingent matter, depending on features of the context.

A contextualist ethic promises to be more effective in identifying and interpreting the moral dimensions of human relations to both domesticated and wild landscapes. Indeed, the very distinction between wild and domesticated turns out to be the product of moral and historical narratives that seek to give meaning to lived experience. The experience of place as wild or domesticated is itself conditioned by the narratives that help construct the meaning of the place and the experience. As Kent Ryden writes, "The sense of place achieves its clearest articulation through narrative, providing the thematic drive and focus from the stories that people tell about the places in their lives."[38] This connection between narrative and the meaning of particular places reinforces the importance of a narrative approach to environmental ethics.[39]

Earlier I argued that the concept of transformative experience was problematic as a device for defending wild nature because what may occasion transformation and the meanings involved in transformative experiences are contextually determined and variable. I have also criticized the anthropocentrism-biocentrism debate because in both instances judgments of value appear to be made by unmarked individuals whose epistemological, let alone geographical, location is undisclosed. A virtue of a narrative ethic approach is that implementing a moral discourse about relationships to the environment requires a situating both of the speaker and of the problem or relationship under discussion.

One example of the blending of moral narrative, moral judgment, and geographical place can be found in Keith Basso's " 'Stalking with Stories': Names, Places, and Moral Narratives among the Western Apaches."[40] According to Basso, the Western Apache at Cibecue in Arizona have incorporated particular features of the landscape into moral narratives. Thus, not only do the narratives make reference to a particular place, but the place itself draws forth reference to the moral narratives with which it is connected. Quoting one source, Basso writes, "The land is always stalking people. The land makes people live right. The land looks after us."[41] Basso's research appears to support Ryden's distinction between a physical landscape and a landscape of meaning formed through narrative: "For those who have developed a sense of place, then, it is as though there is an unseen layer of usage, memory, and significance—an invisible landscape, if you will, of imaginative landmarks—superimposed upon the geographical surface and the two-dimensional map."[42]

According to Jim Cheney, "To get an ethic concerning our obligations within the biotic community, we need to supplement ecological descriptions with narratives told from the inside; we must, as it were, speak *from* our niche, not merely *about* it, in order to get the descriptions we need."[43] In order to judge how we should act in the environment, we must be capable of understanding and articulating the particularities of our relationships with specific environments. Since our judgments presuppose such understandings, we must be prepared to articulate the forces that construct the meanings of nature and to address conflicting and competing interpretations. However, if the contextualist and narrative approach is correct, we can only confront opposing narratives from a situated standpoint, from within the "world" or "form of life" articulated by our own stories. This places a great responsibility on the storyteller not only to listen carefully to the narratives of others, but also to seek conceptual resources for listening critically to one's own.

It might be objected here that Warren's approach substitutes a form of relativism for the abstract universalism of the other positions. Since narrative versions of lived experience frequently conflict, how shall we choose which voice to hear and validate in any given situation?[44] To answer this question, will we not have to fall back on the abstract style of theory to justify the criteria we should use? What, then, becomes of the idea that value is located in relationship, situated in context, rather than in the view from nowhere?

This objection is worth taking seriously. Even if one accepts the notion that moral reality is the product of some form of narrative construction and that there are different standpoints from which moral situations can be interpreted and judged, the moral principle that emerges through the narrative of one person must be attended to and judged by the audience to whom the narration is addressed. While the judgment of the audience may itself take the form of another narrative with its own internal moral perspective, it is nonetheless the case that both narrator and audience articulate ideas about what has value and what does not both in the construction of their narrative and in the response to narrative criticism from others.

There are two responses to this kind of objection. First, the objection calls for abstract, decontextualized criteria of evaluation that the narrative approach suggests do not exist. It calls for a way to judge narratives in advance from a standpoint that is entirely external to the presentation and hearing of narratives by situated individuals in particular contexts. The demand for such criteria begs the question at issue, namely, whether a privileged standpoint external to all particular epistemological and moral inquiries and contexts exists.

Second, the situatedness of storytellers and their narratives need not entail the loss of a critical perspective on narratives and narrative positions. Lorraine Code argues that the form of relativism that emerges from situating knowledge claims and moral judgments in a particular context is compatible with objective and reasoned inquiry.[45] This form of relativism is not the flip side of the universalist view from nowhere.[46] Taylor's "rational and factually enlightened person" is not here being replaced by "anything goes." For the contextualist, consensus about knowledge claims or about moral judgments is a goal to be worked toward, not something guaranteed by the nature of things. Situated knowers and moral judges need not remain isolated by their locatedness, but may participate in extended networks of communication with other situated knowers and judges. The result of mutual and self-critical inquiry may be a collective consensus about what is true and about how we should act. While the contextualist view of objectivity and consensus is precarious, dependent upon the contingent abilities of differently located individuals to communicate, understand, and inquire together, this precariousness need not collapse into a fragmented and alienating relativism. Indeed, one might argue that this view of consensus is no more precarious than that of universalist theorizing that depends upon fictional rational knowers and ideal knowledge conditions to guarantee a consensus.

Thus, Warren's emphasis on the situatedness of the storyteller and his or her lived experience of relationship to nature need not collapse into a cynical relativism that justifies either anthropocentric or biocentric abuses of nature and human beings. Biocentrism as well as anthropocentric theories of human rights, justice, and care may function as tools in the assessment of moral situations, alongside other character traits and cultural norms particular to the situation. But, the relational ethic does not accord these theories any absolute or universal applicability and thus is not compelled to apply their criteria uniformly in disregard for the demands of the situation.

What Warren perhaps does not emphasize enough is the tenuousness of this approach to environmental ethics. There are no built-in grounds for optimism that such a contextualist ethic will achieve consensus on ways of promoting human well-being that respect sustainable relationships with nature. The contextualist avoids ungrounded optimism in the powers of abstract and unsituated rationality to produce legitimacy. The upshot, is a skeptical rejection of the search for closure and finality, which are part of both anthropocentrism and biocentrism.

Philosophical skepticism about theoretical closure and finality seem to entail the rejection of any effort to develop an environmental ethical theory to improve upon anthropocentric theories. But to reject the value of moral theorizing and the search for alternatives to current anthropocentric background beliefs and moral conventions would be too hasty. Indeed, it would neglect the situated needs of our current philosophical and environmental situation. Biocentric environmental ethics serves an intellectual and moral function despite its problems. By articulating alternative conceptions of where moral value can be located, biocentric ethics helps to loosen the self-evidence of anthropocentric assumptions in Western moral traditions. While this conceptual enterprise is flawed when taken abstractly, it is an inevitable accompaniment to that demand of many environmental philosophers for a transformation in our attitudes and emotional orientations to the nonhuman world as an essential precondition of forming sustainable relationships with nature. Kheel may be right that we must care about nature before we can develop an environmental ethic that defends this care, but the formation of caring and sustainable relationships with nature is conceptually linked to the idea that nature has value for its own sake.[47]

Environmental ethics needs both meaningful narratives that particularize and interpret lived experience in relationship to nature as well as the philosophical articulation of grounds for supposing that nonhumans have intrinsic value. Both seem to be essential components of a radical reconstruction of human practices toward nature, their constitutive rules, and their intersubjectively shared meanings and conventions. When taken together in a common critical and reflective project, Warren's relational ethic helps to contextualize the deployment of biocentric arguments, while biocentrism helps, along with other conceptual tools, to articulate aspects of human relationships to nature worth problematizing. Environmental ethical reflection must move along both fronts simultaneously.

I wish to thank Bryan Norton and an anonymous reviewer for their helpful comments on an earlier draft of this article.

Notes

1. Karl-Otto Apel, "The A Priori of Communication and the Foundation of the Humanities," *Man and World*. 5 (Fall 1972): 3–37.

2. Langdon Winner, *The Whale and the Reactor: A Search for Limits in an Age of High Technology* (Chicago: University Of Chicago Press, 1986).

3. Bill Devall and George Sessions, *Deep Ecology: Living as if the Earth Mattered*. (Salt Lake City: Peregrine Books, 1985); J. Baird Callicott, "On the Intrinsic Value of Nonhuman Species," in *In Defense of the Land Ethic: Essays in Environmental Philosophy* (Albany, N. Y. : State University of New York, 1989), 129–156; Tom Regan, "The Nature and Possibility of an Environmental Ethic," *Environmental Ethics*. 3 (Spring 1981): 19–34.

4. Eric Katz, "The Call of the Wild: The Struggle Against Domination and the Technological Fix," *Environmental Ethics*. 14, no. 3 (1992): 271.

5. The anthropocentric claim that only humans and their interests can have moral value should not be confused with the notion that all value is necessarily *anthropogenic*. A biocentrist may agree that all value is created by human valuers, but still argue that nature should be valued for its own sake, in its own right, and not simply for its instrumental value for humans.

6. Doug Aberley, ed., *Futures by Design: The Practice of Ecological Planning* (Philadelphia: New Society Publishers, 1994); David Gordon, ed., *Green Cities: Ecologically Sound Approaches to Urban Space* (Montreal: Black Rose Books, 1990); Charles Hough, *Cities And Natural Process* (London: Routledge, 1995).

7. Paul Shepard, *The Tender Carnivore and the Sacred Game* (New York: Scribner's, 1973).

8. Bryan Norton, "Why I am not a Nonanthropocentrist: Callicott and the Failure of Monistic Inherentism," *Environmental Ethics*. 17, no. 4 (1995): 341–59; Wendell Berry, *The Unsettling of America: Culture and Agriculture* (San Francisco: Sierra Club Books, 1977), 27–30.

9. Bill McKibben, *The End of Nature* (New York: Random House, 1989); William Cronan, *Changes in the Land: Indians, Colonists, and the Ecology of New England* (New York: Hill and Wang, 1983).

10. Donna Haraway, "Situated Knowledges: The Science Question in Feminism and the Privilege of Partial Perspective," in *Simians, Cyborgs, And Women: The Reinvention Of Nature* (New York: Routledge, 1991), 183–202, quote 191.

11. Karen Warren, "The Power and the Promise of Ecological Feminism," *Environmental Ethics* 12, no. 2 (1990): 125–46.
12. William Baxter, *People or Penguins: The Case for Optimal Pollution* (New York: Columbia University Press, 1974), 304–305.
13. Eric Katz and Lauren Oechsli, "Moving Beyond Anthropocentrism: Environmental Ethics, Development, and the Amazon," *Environmental Ethics*. 15, no. 1 (1993): 49–59.
14. Christopher Stone, *Should Trees Have Standing? Toward Legal Rights for Natural Objects* (Portola Valley, Calif. :Tioga Publishing, 1974).
15. Bryan Norton, *Why Preserve Natural Variety?* (Princeton: Princeton University Press, 1987).
16. Norton, *Why Preserve Natural Variety?*, 185–213.
17. Marti Kheel, "The Liberation of Nature: A Circular Affair," *Environmental Ethics* 7 (1985): 339–45.
18. For a more recent discussion of transformative value, see Bryan Norton, *Toward Unity Among Environmentalists* (New York: Oxford University Press, 1991).
19. Carolyn Merchant, *The Death of Nature: Women, Ecology, and the Scientific Revolution* (San Francisco: Harper And Row, 1980).
20. Merchant, *Death of Nature*, 29–41.
21. Paul Taylor, *Respect for Nature: A Theory of Environmental Ethics* (Princeton: Princeton University Press, 1986).
22. Taylor, *Respect for Nature*, 45.
23. Taylor, *Respect for Nature*, 46.
24. Devall and Sessions, *Deep Ecology*, 67.
25. Taylor, *Respect for Nature*, 272.
26. Taylor, *Respect for Nature*, 276–77.
27. William French, "Against Biospherical Egalitarianism," *Environmental Ethics*. 17, no. 1 (1985): 39–58.
28. J. Baird Callicott, "Animal Liberation: A Triangular Affair," in *In Defense of the Land Ethic: Essays in Environmental Philosophy* (Albany, N.Y. : State University of New York, 1989), 15–38.
29. Aldo Leopold, *A Sand County Almanac: With Essays on Conservation from Round River* (New York: Oxford University Press, 1966), 240.
30. Callicott, "Animal Liberation: A Triangular Affair," 34.
31. J. Baird Callicott, "Animal Liberation and Environmental Ethics: Back Together Again," in *In Defense of the Land Ethic: Essays in Environmental Philosophy* (Albany, N. Y. : State University of New York, 1989), 49-59; Mary Midgley, *Animals and Why They Matter* (Athens: University Of Georgia Press, 1983).
32. Callicott, "Animal Liberation and Environmental Ethics," 55.
33. Paul Thompson, *The Spirit of the Soil: Agriculture and Environmental Ethics* (New York: Routledge, 1995); Gary Comstock, "Do Agriculturalists Need a New, an Ecocentric, Ethics?" *Agriculture and Human Values*. 12, no. 1 (1995): 2–16.

34. Ramachandra Guha, "Radical American Environmentalism and Wilderness Preservation: A Third World Perspective," *Environmental Ethics*. 11, no. 1 (1989): 71–83.

35. Warren, "The Power and the Promise of Ecological Feminism"; see also Jim Cheney, "Postmodern Environmental Ethics: Ethics as Bioregional Narrative," in *Postmodern Environmental Ethics*, ed. Max Oelschlager (Albany, N. Y. : State University of New York Press, 1995), 23–42.

36. Warren, "The Power and the Promise of Ecological Feminism," 135–36.

37. Haraway, "Situated Knowledges," 191–93.

38. Kent C. Ryden, *Mapping the Invisible Landscape: Folklore, Writing, and the Sense of Place* (Iowa City: University of Iowa Press, 1993), xiv.

39. Some other treatments of place that are relevant to a contextualist environmental ethics include Eric Hirsch and Michael O'Hanlon, eds., *The Anthropology of Landscape: Perspectives on Place and Space* (Oxford: Clarendon Press, 1995); D. W. Meinig, ed., *The Interpretation of Ordinary Landscapes: Geographical Essays* (New York: Oxford University Press, 1979); Yi-Fu Tuan, *Space and Place: The Perspective of Experience* (Minneapolis: University of Minnesota Press, 1977); and David Seamon, ed., *Dwelling, Seeing, and Designing: Toward a Phenomenological Ecology* (Albany: State University of New York Press, 1993).

40. In *On Nature: Nature, Landscape, and Natural History*, ed. Daniel Halpern (San Francisco: North Point Press, 1987), 95–116.

41. Basso, "Stalking with Stories," in *On Nature: Nature, Landscape, and Natural History*, 95.

42. Basso, "Stalking with Stories," 40.

43. Jim Cheney, "Callicott's Metaphysics of Morals," *Environmental Ethics*. 13, no. 4 (1991): 311–25, quote 321.

44. Roger J. H. King, "Caring About Nature: Feminist Ethics and the Environment," *Hypatia*. 6, no. 1(1991): 75–89, esp. 84.

45. Lorraine Code, "Must a Feminist be a Relativist After All?" in *Rhetorical Spaces: Essays in Gendered Location* (New York: Routledge, 1995), 185–207.

46. Roger J. H. King, "Relativism and Moral Critique," in *The American Constitutional Experiment*, ed. David M. Speak and Creighton Peden (Lewiston, N. Y.: The Edwin Mellen Press, 1991), 145–64.

47. Kheel, "The Liberation of Nature."

Ecology, Modernity, and the Intellectual Legacy of the Frankfurt School

Matthew Gandy

Oil-soaked seabirds lie scattered along a Canadian shore. Children suffering the long-term effects of nuclear radiation lie impassively in a Kiev hospital. Thousands of victims of the Bhopal explosion seek compensation for the impact of industrial negligence. These and many other striking examples of the ecological consequences of modernity lead many environmental thinkers and activists to reject any ideas drawn from the legacy of Western Enlightenment with its anthropocentric stance in relation to nature. But what is the cause of these environmental calamities? Will the abandonment of modernity and the legacy of Western rationalism actually lead us toward a new society in harmony with nature? By abandoning modernity altogether do we risk a chaotic regression wherein ethical principles derived from nature itself undermine any epistemological attempts to understand the cause of environmental problems?

In this paper I draw on the legacy of the Frankfurt School of critical theory in order to argue that ecological problems can best be resolved by an extension of communicative rationality in order to transform relations between society and nature.[1] By communicative rationality I am suggesting that the solution to environmental crisis lies ultimately in the sophistication of our social institutions rather than in the search for new models of society drawn from nature.[2] Central to my argument is the contention that environmental ethics and epistemology are interrelated: the attempt to separate ethics and epistemology may lead to specious appeals to nature-based political ideologies that obscure the capacity for historical change in human societies. My appeal for the development of communicative rationality and a reformulation of the modernity project stems from the normative impossibility of improving social institutions in the absence of a dynamic and critical public sphere. I thus seek to relate environmental discourse to the contemporary processes of social and economic change, which have fundamentally altered the context for policy making in Western societies.

In the first part of the paper I show that there are important areas of overlap between the critique of instrumental reason developed in the postwar writings of the Frankfurt School and the emerging ecological critique of modernity advanced by radical sections of the environmental movement. Second, I develop the concept of communicative rationality in order to show how epistemological issues are central to any discussion of environmental ethics. I argue that the Habermasian conception of communicative rationality contains a number of important weaknesses born out of its unnecessarily restrictive ethical and epistemological stance with respect to nature. Finally, I explore the current attempt to construct an ecological Enlightenment around a revitalization of the modernity project advanced by Ulrich Beck. I argue that this marks an important advance on the earlier treatment of ecological issues by the Frankfurt School, yet the normative value of Beck's analysis is hampered by an overemphasis on the technological dimensions to social change. I explore this weakness by drawing on an alternative reading of the tension between ecology and modernity advanced by the North American inheritors of the early Frankfurt School tradition.

Modernity in Question: The Critique of Instrumental Reason

One of the first major engagements with relations between society and nature in the writings of the Frankfurt School is contained in Theodor Adorno and Max Horkheimer's *Dialectic of Enlightenment* published in 1947.[3] For Adorno and Horkheimer, concern with the mastery and destruction of nature forms a central element in their critique of instrumental reason as the shadowy side to Western Enlightenment. Emphasis is placed on science and technology as an advancing system with an internal dynamic and logic of its own, dangerously adrift of civil society. This concern with the ecological consequences of modernity—the destruction of natural beauty, the restriction of public debate to narrowly-defined technical criteria, and above all, the notion of a radical discontinuity between the rhetoric of progress and the reality of social and environmental disarray in the twentieth century, suggests a shared perspective between many of the central tenets of critical theory and radical ideas developed in the environmental literature. I argue in this paper, however, that the interrelation between the insights of critical theory and the radical environmentalist critique of modernity is far more complex and problematic than is widely recognized.

Recent years have seen a growing interest in critical theory from the more theoretically inclined environmental literature. One of the most significant recent responses to the work of Adorno and Horkheimer is provided by Robyn Eckersley. She explores the critique of instrumental reason and the extent to which this might lay the basis for an alternative environmental ethic. For Eckersley, the particular significance of critical theory lies in the shift away from the economic determinism of orthodox Marxism, thereby allowing emphasis on different sources of oppression and exploitation in capitalist society. She identifies parallels between the work of Adorno and Horkheimer and "the ecological critique of industrial society" as it has emerged since the 1960s.[4] Yet her ecocentrist standpoint tends to overemphasize the technological dimension to modernity and elide the critique of instrumental reason with a complete rejection of Enlightenment. Her reading of the critical theory literature views the work of Adorno and Horkheimer through the binary and simplistic analytical framework of ecocentrist and technocentrist thought, a distinction that blurs rather than illuminates our understanding of the environmental crisis.

Adorno and Horkheimer present us with one of the most sustained and influential critiques of Enlightenment rationality yet written but their legacy raises numerous unresolved questions concerning the role of science and technology in an "incomplete" modernity, themes that are tackled in greater detail in the later work of Herbert Marcuse, Jürgen Habermas, and Ulrich Beck. In the writings of Herbert Marcuse, the earlier focus of the Frankfurt School on positivist science is extended into a generalized indictment of a manipulated mass society shackled to ever-greater material consumption. Marcuse develops Adorno and Horkheimer's theme of the "totally administered world" and "the end of the individual" embodied in the decline of subjectivity.[5] Marcuse shows how the technological question emerges as one of the central contradictions of twentieth-century modernity through its simultaneous transformation of both external and inner (human) nature. In *One-Dimensional Man* (1964), he draws together the ideas of Edmund Husserl, Gaston Bachelard, Freud, Marx, and other radical thinkers, into a provocative synthesis to expose the decline of freedom and creativity in the ostensibly stable postwar era.[6] For Marcuse, it is the remarkable continuities and similarities between fascism, capitalism, and state socialism that are striking as the technological and scientific problems of modernity appear to pervade all of these political systems. His emphasis on the increasingly dominant role of science and technology can be illustrated by the following passage:

> The principles of modern science were *a priori* structured in such a way that they could serve as conceptual instruments for a universe of self-propelling, productive control; theoretical operationalism came to correspond to practical operationalism. The scientific method which led to the ever-more-effective domination of nature thus came to provide the pure concepts as well as the instrumentalities for the ever-more-effective domination of man by man *through* the domination of nature. Theoretical reason, remaining pure and neutral, entered into the service of practical reason. The merger proved beneficial to both. Today, domination perpetuates and extends itself not only through technology but *as* technology, and the latter provides the great legitimation of the expanding political power, which absorbs all spheres of culture.[7]

We can identify a series of themes here central to the treatment of nature in critical theory: the interrelationship between the treating of both people and nature as mere instruments of destructive productivity; the mask of ethical neutrality behind which positivist science and technology extend their influence and control; the blurring of distinctions between scientific and practical reason (a tension to be extensively explored in the work of Habermas); and finally, the service of technology to capital accumulation, as all potential sources of human creativity and criticism are subsumed within consumer culture both for the creation of new markets and to extinguish any potential sources of opposition.[8]

There is clearly a tension in the work of the Frankfurt School between the idea that it is scientific epistemology itself that lies behind the destruction of nature and the differing view that the problem stems from the misapplication of science and technology. In this sense, the ideas of Marcuse lie much closer to radical ecocentrist strands of environmentalist thought than the subsequent writings of Habermas, Beck, and Feenberg, with their emphasis on the possibilities for rational discourse in order to democratize the applications of science rather than the abandonment of existing epistemological approaches. The Marcusian perspective finds resonance with a number of ecological thinkers drawn from the post-Marxist left such as Murray Bookchin and André Gorz. Yet Marcuse's emphasis on human emancipation through ecological politics distances him from the more recent development of ecocentrist and deep ecology literature, which draws its inspiration not from radical humanism but from nature-based philosophies. The careful delineation of the boundary between the natural and the social in the work of Marcuse prefigures the concerns of Habermas with the distinction between the empirical-analytical systems of the physical and biological sciences and the hermeneutic realm of the lifeworld rooted in language.

Communicative Rationality and Environmental Ethics

In order to develop our understanding of the potential role of communicative rationality in the development of environmental ethics, I want to begin by exploring the avowedly anthropocentric stance of Jürgen Habermas. At an epistemological level, Habermas does not seek either a reconciliation with nature or a new approach to science. He envisages a dualistic distinction between instrumental rationality governing the nonhuman world of biophysical systems and communicative rationality governing social relations in the lifeworld. Habermas rejects Marcuse's view of science as predicated on a confusion between the objective manipulation of nature by labor and technology to satisfy human needs and the symbolic communicative sphere rooted in language. He therefore rejects the Marcusian contention that society cannot change without a transformation in science and technology. Habermas sees the scientization of politics as the transformation of practical reason (moral-political questions) into instrumental reason (technical questions).[9] For Habermas, the increasing scientization of politics over the modern period has led to the restriction of a deliberative democratic public sphere and the transfer of politics to technical and administrative elites promoting technical rather than political solutions to human problems. He is thus cautious of any naturalistic framework that elides fundamental differences between biophysical systems necessary for survival and socio-cultural communicative systems necessary for historical progression. This premise is based on his distinction between the natural sciences and the historical-hermeneutic sciences, between technical control (instrumental reason) and understanding (practical reason), in order to emphasize how the goal of social communication should be different from our relations with the natural world.

Habermas is clearly a defender of the modernity project and the changes in human relations with nature that this entails. He rejects the utopian aesthetic concerns of Adorno for the re-enchantment of the social and natural worlds, and moves the focus of his analysis to questions of social justice.[10] Habermas is not so much concerned with the instrumental reason of science under modernity as with the lagging development of the communicative and democratic public sphere.[11] The potential contribution of Habermas to environmental thought stems principally from his concerns with the scientization of the political process and the attempted depoliticization of the public sphere. The scientizing of environmental discourse can be seen as an attempted technical resolution of crises in the public sphere stemming from deep seated contradictions between democracy and modernity under the administrative apparatus of the state. An informed citizenry is increasingly marginalized in relation to an array of technical experts across diverse fields of concern ranging from the promotion of nuclear energy to the release of genetically modified organisms into the environment.[12] He presents us with a post-Cartesian epistemology distinguishable from both scientism and metaphysics, thus placing him apart from positivist technocratic strands of environmental thought and nature-based sources of understanding.

Given this philosophical stance, it is not surprising that the relationship between Habermas and the new social movements has been fraught with difficulty. He has, with the exception of feminism, characterized the ecology and antinuclear movements as defensive rather than emancipatory, as indicative of legitimation problems and the colonization of the lifeworld in advanced capitalist societies.[13] For Habermas, these counterinstitutional struggles from within the lifeworld are futile without any transformation of the structure of society:

> neo-populist protests only bring to expression in pointed fashion a widespread fear regarding the destruction of the urban and natural environment, and of forms of human sociability. There is a certain irony about these protests in terms of neoconservatism. The task of passing on a cultural tradition, of social integration, and of socialization require the adherence to a criterion of communicative rationalization occasions for protest and discontent (which) originate exactly when communicative action, centred on the reproduction and transmission of values and

norms, are penetrated by a form of modernization guided by standards of economic and administrative rationality; however, those very spheres are dependent on quite different standards of rationalization—on the standards of what I would call communicative rationality.[14]

The environmental movement for Habermas is thus deeply contradictory in its relationship toward the outcome of modernity. It is on the one hand critical of the social and economic relationships that have facilitated its own emergence yet it remains unable to see beyond the circumstances of a partial modernity with its unequal penetration of rational discourse in public life. This contradiction is perhaps most intense in the field of risk where positivist scientific epistemologies are routinely used by the environmental movement in order to expose the epidemiological basis of threats to public health in conjunction with broader appeals against modernity *tout court*. If we disentangle environmental discourse, we find a complex medley of ethical and epistemological issues nowhere more confused than in the ecocentrist appeal to nature as a privileged source of invariant meaning. We can find many examples in the environmental literature where the critique of science and technology is conjoined with an ideologically ambiguous political agenda drawn from nature itself.[15] In other cases the science of ecology is extended to encompass all aspects of human society in an ahistorical socio-biological system amenable to control by an array of experts. In recent years, this "cybernetic hyperscientism" has been able to gain a degree of credibility through the manipulation of huge databases in sophisticated models of biophysical systems far in advance of the earlier attempts of the 1970s.[16]

The idea of ecology as a holistic metascience, a new "grand theory" based around "technocratic and dystopian fantasies of total administration" capable of directing human thought and action, remains influential across a wide spectrum of environmental thought, yet these ecologically based social theories rest on an overextension of the epistemologies of the natural sciences.[17] Habermas is skeptical of the possibility for rethinking the human relationship to nature because this entails a breakdown in rational scientific discourse and a retreat into metaphysics beyond the reach of the empirical-analytic sciences.[18] The tension between Habermas and the Green movement can be traced to his theoretical break from the negative dialectics of Adorno and Horkheimer and their goal of "reconciliation with nature." For Habermas, the idea of nature-in-itself is necessarily trapped within a transcendental framework and the search for ultimate origins: a Heideggerian project rooted in a search for primordial and mythical ontologies with disturbing political implications for rational discourse.[19]

If Habermas is at such pains to formulate a version of communicative rationality that excludes any appeal to nonhuman nature, then what are the implications for environmental ethics? What might a nonpositivist communicative rationality actually mean for normative environmental discourse? If we take as a starting point the promotion of the public interest in the arena of health and quality of life, then communicative rationality can be argued to endorse a social and economic system that allows both the promotion of social justice and long-term environmental sustainability since human well-being and environmental quality can be demonstrated to be interrelated. This is in essence an extended notion of enlightened self-interest where an environmental ethic can be established without appeal to intrinsic values in nature. The argument that an underlying rational harmony exists between the interests of human and nonhuman nature now forms a key dimension to the debate over the role of humanism in environmental ethics.[20] In contrast, deep ecological perspectives have sought to dispense with humanism completely because of the implicit anthropocentric and utilitarian impulses. Questions concerning the sentience of nonhuman nature and the boundary of moral considerability have been extended by biocentric ethics to include trees, rivers, and geomorphological features.[21] The influential writings of Arne Naess and Warwick Fox, for example, lead us toward the normative and epistemological weaknesses of nature-based philosophical discourses that illuminate the concerns of Habermas with metaphysical ontologies.

It is certainly the case, however, that a strict epistemological division between society and nature may serve to exclude the ethical handling of nonhuman nature, rendering the protection of species and ecosystems that have no direct instrumental value problematic.[22] The relegation of nature to the empirical-analytic sciences in Habermas's philosophical schema necessarily excludes human relations with nature from the historical-hermeneutic sciences and ultimately fails to challenge narrowly instrumentalist views toward nature. His conception of ethics combines a contractual utilitarian dimension with a Kantian concern for intrinsic rights but does not extend to nonhuman nature.[23] The advent of environmental ethics and the extension of ethical consideration to nonsentient living things and ecosystems poses a fundamental challenge to existing utilitarian, Kantian, and contractarian views of ethics.[24] In the place of the focus on the individual and his or her relations with others in society emerges an expanded notion of the self to encompass a wider biotic community of which we are a part. It is, however, the ontological limits to rational discourse that ultimately place the more radical variants of environmental ethics beyond the practical concerns of critical theory.

I want to place the ethical dimension to communicative rationality to one side for a moment in order to explore more closely the epistemological weaknesses in Habermas's conception of relations between society and nature. The possibilities for critical analysis of phenomena that lie at the boundary of natural and social systems are unnecessarily weakened by his strict demarcation of human and nonhuman forms of social interaction. This issue is raised by Bruno Latour, who finds the epistemological barrier to nature in Habermas's work to be nonsensical since these divisions are becoming increasingly blurred with every advance in medicine and genetics. There is a proliferation of "quasi-objects" entering the world, which are neither wholly natural or social:

> If anyone has ever picked the wrong enemy, it is surely this displaced twentieth-century Kantianism that attempts to widen the abyss between objects known by the subject on the one hand, and communicational reason on the other. . . . Habermas wants to make the two poles incommensurable, at the very moment when quasi-objects are multiplying to such an extent that it appears impossible to find a single one that more or less resembles a free speaking subject or a reified natural object.[25]

However, to focus our attention exclusively on the problematic way Habermas handles nature risks overlooking a central focus of his work: the defense of the modernity project against the rival post-Hegelian discourses of poststructuralism. Habermas is concerned to show how modernity can critically reassure itself as to its rationality in distinction to the "total critique of reason" drawn from the legacy of Nietzsche, Bataille, and Foucault and the metaphysical turn embodied in the writings of Heidegger and Derrida.[26] This brings us back to my central concern in this paper: the extent to which the ecological problems of modern societies can be solved by an extension of communicative rationality within society without necessitating new forms of interaction with, and understanding of, nature itself.

The most incisive contribution of Habermas to the environmental debate is his concern with the public sphere. He opens his major treatise on the topic with direct reference to the centrality of the idea of "the public" in the development of Western thought and to its continued relevance to any understanding of contemporary political debate:

> The usage of the words "public" and "public sphere" betrays a multiplicity of concurrent meanings. Their origins go back to various historical phases and, when applied synchronically to the conditions of a bourgeois society that is industrially advanced and constituted as a social-welfare state, they fuse into a clouded amalgam. Yet the very conditions that make the inherited language seem inappropriate appear to require these words, however confused their employment.[27]

During the course of this work, Habermas explores the transformation of a critical public sphere into a manipulated consumer society and the concomitant weakening of public institutions vested in civil society. We can argue that the Habermasian ideal of some form of rational universal consensus is more conducive to ecological sustainability than individualist liberalism with its weakly defined public realm which has been so easily eroded under neoliberal public policy. Perhaps the most prominent exponent of these insights for environmental discourse is John Dryzek, who defines the public sphere as "the space in which individuals enter into discourse that involves mutual respect, openness, scrutiny of their relationship with one another, the creation of truly public opinion, and, crucially, confrontation with state power."[28] But what kind of public sphere for what kind of public policy is implied here? The Habermasian notion of a public sphere has been criticized from a number of quarters: by Marxists in terms of class, by feminists in terms of gender, by neoliberals in terms of ignoring private interests, and by poststructuralists for putting forward the very idea of a universal social consensus.[29] Thus Habermas is vulnerable to the charge of presenting a quasi-scientific, Eurocentric, and androcentric ideological position under the guise of universality and communicative solidarity.[30] The notion of a public interest expressed through a socially mediated response to environmental concerns is clearly more complicated than the Habermasian conception of communicative rationality will allow. In the absence of some kind of broad-based agreement over the truthfulness of scientific claims, it is difficult to see how public policy can effectively operate in the environmental arena. My contention in this paper is that in the absence of an appropriate forum or mechanism for reaching agreement over the ends and means of public policy we are left with an increasingly market-led technocratic approach to environmental management where those concerns that do not readily contribute to capital accumulation or the quantitative logic of risk assessment and cost-benefit analysis will be eclipsed from environmental discourse. The centrality of the public sphere to environmental concerns stems from the practical need to find agreement over the extent of ecological problems and to develop the intricate social arrangements necessary for their resolution. Every human society both modern and premodern has developed ways of handling its relations to nature but under capitalist urbanization and the globalization of modernity the relationship has become increasingly difficult to sustain.

The Habermasian public realm reveals an underlying tension between ecology and modernity over which critical theory and radical environmentalism part company. Whereas Habermas seeks to defend and elaborate the Enlightenment project, Eckersley and the ecocentrists see the mastery of nature under modernity as an illusory and undesirable goal that denies the interdependence between social and biophysical systems.[31] Yet Eckersley's critique of critical theory is undermined by the weakness of her exploration of the tension between ecology and modernity and her insistence that instrumental reason is the primary cause of environmental destruction. Tim Hayward quite rightly points out that it is only by appreciating the contemporary divergence between Habermasian and poststructuralist readings of modernity that we can appreciate the underlying tensions between ecology and modernity obscured in Eckersley's naively "ecologistic" reading.[32] To question the rationality of positivist science and technocratic reason is not, therefore, to suggest the redundancy of the Enlightenment ideal of reason altogether. The most important question to emerge from the work of Habermas is whether there can be an ecological rationality derived from the full development of the communicative and democratic dimensions to social life. His faith in developing a communicative realm capable of handling developments in science and technology in the public interest rests ultimately on an anthropocentric vantage point in the interests of epistemological rationality. The work of Habermas is clearly distinguishable from two competing discourses on nature: first, that of technical mastery under the instrumental reason of positivist science; and second, the metaphysical irrationalist strand linking nineteenth-century romanticism with contemporary ecocentrist and deep ecological formulations. In the final analysis, however, Habermas's conception of communicative rationality remains insufficiently developed to realize its potential role in the development of an environmental ethic within a reformulated modernity. His notion of a public sphere is too restrictive in relation to both nature and social difference to take account of the extensive interweaving of nature and culture in contemporary society.

Ecology and the Reconfiguration of Modernity

I want to turn now to the contemporary handling of the ecological question in critical theory and draw on the pathbreaking work of Ulrich Beck. Though Beck cannot properly be placed within the central canon of critical theory, his work represents in many ways a logical extension and development of themes prefigured in the earlier writers of the Frankfurt School. Of particular interest here is Beck's elaboration of the critique of scientism and his challenge to the "organized irresponsibility" of contemporary society.[33] He draws attention to critical new developments such as the human genome project where life itself is now under greater technical mastery than anything envisaged in the time of Adorno or Horkheimer. Beck warns of a process of "eugenics by stealth," as genetics may ultimately displace social policy as an interrelated nexus of technical mastery and control. The question of rationality is thus brought to center stage, since "it is not an excess of rationality, but a shocking lack of rationality, the prevailing irrationality, which explains the ailment of industrial modernity. It can be cured, if at all, not by a retreat but only by a radicalization of rationality, which will absorb the repressed uncertainty."[34] Beck is thus concerned with the ecological consequences of modernity but suggests that we must work from within the modernity project itself. In this respect we can distinguish Beck from the ecocentrist environmentalists who argue for an abandonment of Enlightenment rationality in order to tackle the environmental crisis. As in the case of Habermas, we can differentiate between Beck's insistence on the need for a normative rationality and the ecocentrist search for innate sources of meaning residing within nature itself. Unlike Habermas, however, Beck seeks to abandon the increasingly false dichotomy between nature and society, which pervades so much of the environmental literature.

In *Risk Society*, first published in 1986, Beck shows how the productivist logic of industrial modernity systematically neglects and ignores associated risks from sources such as nuclear technologies, genetic engineering, manufacture of toxins, and climate change. He elaborates the metaphor of risk to show how society is now confronted by itself under a condition of "reflexive modernity" in contrast to the earlier largely external sources of risk prevalent in premodern societies: "the sources of danger are no longer ignorance but *knowledge*; not a deficient but a perfected mastery over nature. . . . Modernity has taken over the role of its counterpart—the tradition to be overcome, the natural constraint to be mastered. It has become the threat *and* the promise of emancipation from the threat it creates itself."[35]

For Beck, the transformation of the political process in risk society has several interrelated elements. First, there is a disjunction between the processes of societal transformation and the restricted arenas of political discourse, marked by a crisis of governance in existing political institutions. In the sphere of science and technology, the increasing severity of risk undermines rationality in public policy leading to a widening gap between state authority and the democratic awareness (and expectations) of citizens. This disintegration of politics is marked by the declining legitimacy of state intervention and occurs in the midst of a growing political challenge to scientism and technological rationality. Thus the unraveling of any harmony between social and technological progress emerges as a central theme in Beck's risk society and is fundamental to the rise of the new social movements with their destabilizing impact on the postwar consensus.

Second, as an outcome of these changing relationships there is a reversal of political and nonpolitical realms, as the relative disempowerment of the state and established areas of public policy is accompanied by the extension of the political process into what Beck refers to as "the sub-political system of scientific, technological and economic modernization."[36] The locus of political power shifts decisively from the state and political parties to the boardroom, the research laboratory and the grassroots arena of "sub-politics," thus unraveling the administrative dimensions to the "one-dimensional society" of the postwar era and signaling the emergence of a society ever more remote from the classic conception of a liberal public sphere within which civil society has the opportunity to deliberate over matters of public concern. Phenomena of increasing ungovernability and the hollowing out of the state lie in juxtaposition with the increasing severity and complexity of the social and ecological consequences of late capitalism. Uncertainty emerges as the political and cultural counterpart to economic flexibilization in the post-Fordist era. Existing patterns of interest and political alliances are placed in a state of flux as the distribution of winners and losers shifts in industrial risk society: the ecological contradictions of capital become ever more intense as the "invisible hand" becomes an "invisible saboteur" of investments and profits.[37]

This inherent instability of risk is intensified under pressures for greater environmental deregulation. Consider the example of the spread of BSE (bovine spongiform encephalopathy), better known as "mad cow disease," through the British beef industry over a period of some sixteen years due to the deregulation of intensive food production. The outcome has been a dramatic disarray in U.K. agriculture, with the interests of producers and consumers brought into direct conflict with each other. In the face of increasing numbers of human deaths from a BSE-related form of CJD (Creutzfeldt-Jakob Disease) attributed to infected meat, the parameters of uncertainty and risk now extend to a public health epidemic in combination with fiscal chaos arising from the mass slaughter of cattle and loss of trade. The issues of rationality in public policy and the regulation of technological developments are brought to center stage as beef consumption throughout Europe has been adversely affected. We are faced with a public health crisis that is simultaneously derived from the modification of nature by human agency, the market-led deregulation and restructuring of food production, and the social and cultural responses to risk. At an epistemological level we are in a realm that cannot be neatly demarcated in the Habermasian sense, nor left to the exigencies of positivist risk assessment and relativist cultural constructivism. The extensive public disagreement among scientists and experts underlies the problematic status of "truth" in environmental discourse and exposes the current inadequacy of the scope for public deliberation and understanding of these issues.

For Beck, the increased questioning of scientific truth claims since the 1970s stemming from concern over the deleterious impact of science and technology is in effect a radicalized challenge to positivist science in order to build a more defensible scientific rationality. In the place of scientific certainty emerges explicit recognition of the socially negotiated dimensions to truth with far-reaching implications for policy making and the relationship between science and society. Beck articulates a postpositivist perspective on science and technology in dialectical relation to itself, where the internal divisions and disarray of experts within science become ever more advanced. He provides a post-Chernobyl critique of instrumental reason developed to encompass the consequences of increased risk in relation to the narrowly defined notions of modern rationality and self-interest:

Political development in hazard civilization is approaching the crucial issue of the redistribution and democratic shaping of the principles, rules and foundations of the power to define terms: different relations of proof, different relations of restraint, different relations of control and guidance, different relations of participation in decision-making. . . . The character of industrial society is such that its momentum contradicts self-determination, as fatalism contradicts democracy, and organized non-liability contradicts rationality and justice.[38]

Beck is suspicious of any drift towards an ecological welfare state because of the persistence of antidemocratic tendencies inherent in centralized administrative structures. Indeed, he sees the declining relevance and legitimacy of the state as a fundamental dimension to the "incomplete modernization" of society. Yet this is not the antistatist sentiment of right-wing ecologism but rather the recognition that a completely new regulatory regime is vital, within which the role of law is crucial in mediating between members of an increasingly heterogeneous society in the absence of mass political parties and clearly defined programmatic agendas for change. Beck echoes the desire of Habermas for the creation of conditions in civil society within which a rational consensus through democratic deliberation can be reached. As Michael Rustin puts it, "Beck evokes the possibility of a fully conscious, rational society, able to take full responsibility for its development and for its relationship with nature."[39] Full modernity is therefore conceived as a condition that has not yet arrived and should not be confused with transitional and much-maligned phases of social development such as the high modernity of Fordist technocracy or the high-rise housing fiascos associated with the International Style in architecture. It is only under full modernity that both the natural and social worlds can be brought "within the spheres of understanding and choice," and allow the world to be shaped by human reason.[40]

The concepts of "risk society" and "reflexive modernization" may appear superficially persuasive. But is Beck's conception of "full modernity" naively at odds with current patterns of social and economic change? Beck's writings display a tendency toward a pluralist view of political conflict in his lack of acknowledgment of systematic inequalities in the distribution of power between different institutions. This is related to a micro-political bias and a tendency toward an individualized conception of social processes suggested by his contention that "the microcosm of daily behaviour and dealings with oneself and others corresponds to the macrocosm of threat production."[41]

Likewise, his writings on the "metamorphosis of the state" suggest that the main alternative centers of power lie in the proliferation of self-organized interest groups at the "sub-political" level.[42] Though Beck rightly admonishes simplistic *post histoire* conceptions of society as having reached "the end of history," his work does display a tendency toward a teleological and technologically driven notion of "post-industrial society" predicated on his extensively employed contrast between "classical risk society" and "industrial risk society."

In arguing that the "compulsory union of industrial society and modernity can be broken," Beck is implicitly restricting the spatial scope of his analysis to technological developments in the core economies of the West, thereby overlooking the spatial restructuring of economic production at a global level.[43] Beck calls for "the totality of bureaucratic-industrial-political supremacy" to be placed at the center of an oppositional politics but never demonstrates how such a realignment in political conflict might occur or how it would alter the trajectory of social and economic development. There is an all-pervading focus on what is variously referred to as "techno-scientific rationality" or "technocracy" rather than on the institutional power of capital, thus overemphasizing the ideological strength of science and technology in relation to the cultural hegemony of capital. Yet this concept of "techno-science" in environmental discourse blurs the distinction between the pervasive use of new technologies in everyday life and the relatively hidden realm of scientific research. In the writings of Andrew Feenberg, for example, we find a reworking of the utopian sentiments of the early Frankfurt School rooted in a critique of the role of capital in distorting the potentially liberatory role of technology in society.[44]

A similar theme is developed by William Leiss, another inheritor of the early Frankfurt School tradition, who emphasizes the need to resist a fatalistic "technological fetishism," which undermines the need to make reasoned choices about societal development.[45] Leiss describes how Western "scientific culture" acts as a powerful ideological link between the natural sciences and popular aspirations for material well-being. Yet this "techno-scientific" material promise of a better life is predicated on a universalization of dominion over nature and the generalized wastage of materials and resources.[46] Leiss invokes a Marcusian concern with the recovery of the sensual side to nature and human well-being yet is careful to resist any drift into nature-based rationalizations of social relations. It is at the political level, however, that the starkest differences between Beck's technological preoccupations and these alternative readings of critical theory become most clearly apparent.

The downplaying of the role of capital in Beck's analysis undermines his handling of the state and environmental regulation. In relegating the welfare state to the discarded baggage of industrial modernity, he exhibits a certain ambiguity with regard to socioeconomic restructuring and a reluctance to acknowledge both the primacy of the needs of capital in this process and, conversely, the undermining of the public realm in key areas such as education, research, health, and environmental protection. In other words, he has little to say about what the postwelfarist world might look like and whether it significantly advances his goal of a new modernity. Beck neglects to consider whether the scientization of politics can be conceived as part of a broader process of state restructuring where government functions are removed from the democratic arena in order to facilitate greater fiscal control over expenditure. He pays scant regard to the regulatory structures that are needed to tackle complex international problems such as the export of toxic waste, nuclear proliferation, and climate change and neglects to examine the historical dimensions to environmental regulation in any detail. Beck's widely repeated dictum that "poverty is hierarchical, while smog is democratic" ignores wide and growing sociospatial inequalities in environmental quality.[47] These have been extensively exposed by the growth of the environmental justice movement in the 1990s in opposition to the "toxic industrial spaces" of late capitalism arising from the increasing spatial concentration of environmental externalities under the postwar legislative drive of environmental regulation.

Consider, for example, the intense concentration of polluting industries along the so-called Cancer Alley between New Orleans and Baton Rouge, where poor communities have been faced with high incidences of cancer, birth defects, and miscarriages.[48] Beck's assertion that industrialization means both wealth *and* proximity to industrial hazards clearly oversimplifies the spatial distribution of risk.[49] With the emergence of radical environmental groups across the United States such as the Gulf Coast Tenants Leadership Association, the Mothers of East Los Angeles, and the South Bronx Clean Air Coalition, the contours of power in corporate decision making and state regulation become exposed as the sites of resistance to environmental degradation in capitalist society. The inequalities by race, class, gender and age that determine exposure to risk and poor environmental quality are brought to the fore in environmental discourse. In Beck's "industrial risk society," however, it is difficult to discern where conflicts of interest lie since his work tends to downplay the significance of grassroots political struggles in the workplace or community in favor of a more abstract focus on fear and doubt in society as a whole. In arguing that "the tradition of intervention and resistance has wasted away," he is subsuming the demise of Marxist thought with the disappearance of political activity in his own *post histoire* account of contemporary social change.[50]

The most important insights to emerge from the work of Beck stem from his recognition of the fundamentally altered nature of the relationship between society and nature under late modernity. Beck's work provides an important advance on Habermas in that he recognizes the need to find an epistemological middle ground between what he terms the scientific and technical recognition of environmental threats and the "cultural and symbolic mediation of the consciousness of threat."[51] Beck presents us with a plea for an environmental ethic where the tension between rationality and irrationality is embodied in a recognition of both the consequences and opportunities of modernity. The unresolved question to emerge from his analysis is how the pervasive sense of doubt under the irrationality of reflexive modernization can be transformed into an "ecological Enlightenment" in the face of sustained opposition from the powerful interests who have benefited from the unequal distribution of risk.

Conclusion

The dominant tension running through Western intellectual debates since the early 1980s has been between the possibilities for a reformulated modernity and a complete abandonment of the Enlightenment project. If we reconsider this hiatus from an ecological perspective, there is a clear distinction between the ecocentrist rejection of modernity and the Habermasian concern with the relative imbalance between the realms of science and communicative rationality. Though there are epistemological weaknesses in relying on a Habermasian conception of nature, this does not detract from the normative significance of his writings on the public sphere. All the most significant and important advances in relations between society and nature have been rooted in the articulation of a public good above and beyond any narrowly conceived notions of self interest. Perhaps the ultimate paradox of the tensions between ecology and modernity is that while one can point to many very real achievements (I am thinking here especially of advances in the fields of medicine and public health), the overall trajectory of social change in recent decades has been overwhelmingly inimical to the long-term stability of life sustaining biophysical systems. This suggests that in any meaningful discussion of the relations between ecology and modernity we need to clarify the contradictory and diverse impulses within modernity itself in order to distinguish between its constituent elements as they span across science, technology, capital, and ideology. This involves the recovery of a modernity rooted in the realization of human potential and the affirmation of life over death: a radical reworking of the discourses of "nature" to affirm the sensuality and pleasures of existence.[52] At a political level, such a project necessarily demands a disengagement of human satisfaction from the technomilitary complexes that sustain the vapid inducements of consumer capitalism.[53] An ecological agenda that refuses to engage with the crisis of modernity ignores the very forces that propel the possibilities for change: there is no way back to the illusory space and time of a premodern "golden age."

If contemporary critical theory is essentially concerned with establishing the basis for a reformulated modernity, then where does this leave the question of political praxis in environmental discourse? Radical environmentalists have charged critical theorists with providing an overly abstract model of society, within which an anthropocentric instrumental reason is perpetuated. Although there is certainly a strong case for the reexamination of environmental ethics for the treatment of nonhuman nature within critical theory, this does not justify an abandonment of any emphasis on the need for a revitalized public realm. If we reexamine the main tension within environmental ethics between anthropocentric and biocentric conceptions of relations with nature, we find that critical theory opens up the possibility of an environmental ethic that remains rooted in social practice yet enables a critical perspective on technocratic attitudes toward nature through the critique of positivism, scientism, and instrumental reason. Yet our understanding of the relationship between communicative rationality and environmental ethics leaves many unresolved questions. How, for example, can any consensus over ecological rationality be reached in the face of not only the ideological power of consumer culture but also the increasing individualization and globalization of society? How can the socialization of nature be epistemologically handled in order to provide a degree of normative adequacy for environmental ethics? Does ecological rationality imply little more than a more sophisticated variant of "enlightened self-interest" in environmental ethics? Or can we conceive of ecological rationality as a fundamental challenge to mainstream environmental discourse?

It is undoubtedly the case that critical theory has had less impact on environmental debate than one might expect given the scope of its intellectual heritage. I would suggest that part of the reason for this hiatus between theory and praxis concerns the problematic status of post-Marxist theory within the context of a widening gap between the radical academy and the rest of society. A body of ideas that combines abstract thought in the structuralist tradition with a total critique of existing society is difficult to reconcile with the demands of positivist "relevance" in academic research or the relativist disdain for normative theorizing. One hopes, however, that the potential form of a new modernity will be forged as much in the world of ideas as in the realm of practical action. It is perhaps only through the stubborn refusal of this legacy of intellectual thought to accommodate itself either to academic fashion or to neoliberal *zeitgeist* that the enduring insights of this tradition will contribute to the difficult work of delineating the epistemological and ethical basis of a modernity freed from both the false claims of positivist science and the search for "new certainties" in nature.

Acknowledgments

I would like to thank the anonymous referees [Andrew Light and Neil Smith] for their detailed comments on an earlier draft of this paper. The funding of sabbatical leave was provided under the Global Environmental Change program of the Economic and Social Research Council.

Notes

1. In order to clarify the scope of this paper, I should set out what I mean by critical theory. We can distinguish between two main uses of the term: first, to refer specifically to the work of writers associated with the Frankfurt Institute for Social Research; and second, in a broader sense to a wide-ranging European Marxist tradition placing greater emphasis on cultural and aesthetic issues than in Marx's original writings. A primary concern of the Frankfurt School has been the interdisciplinary extension of Marxism in conjunction with a critical response to the changing circumstances of capitalist society for

Ecology, Modernity, and the Frankfurt School 251

social and political transformation. Central to this aim has been the attempted reconciliation between Western philosophical traditions and new advances in the empirical sciences. For general overviews of the genesis of the Frankfurt School see M. Jay, *The Dialectical Imagination: A History of the Frankfurt School and the Institute of Social Research 1923–50* (London: Heinemann, 1973); S. Buck-Morss, *The Origin of Negative Dialectics* (New York: Free Press, 1977); T. Bottomore, *The Frankfurt School* (Ellis Horwood: Chichester, 1984); A. Feenberg, *Lukacs, Marx and the Sources of Critical Theory* (Oxford: Oxford University Press, 1986); R. Wiggerhaus, *Die Frankfurter Schule* (Munich: Hanser, 1986); and D. Kellner, *Critical Theory, Marxism, and Modernity* (Cambridge and Baltimore: Polity and Johns Hopkins University Press, 1989).

2. The term "communicative rationality" requires some clarification. I use the word "rationality" here to refer to social practice rather than to specific forms of logic or cognition. When the term is used in reference to social practice, we can distinguish between restrictive uses based around ideas of self-interest such as "rational economic man" and more complex applications rooted in linguistic communication where there is an incorporation of ethical or moral dimensions to judgment. The emphasis here is on rationality as open to negotiation and historically constructed rather than an innate determinant of human interaction in atomized and ahistorical conceptions of society. Ecological rationality is especially complex because it combines a series of tensions between individuals, society, and nature mediated by social difference and intersubjective understanding. For the purposes of this paper, I restrict my discussion of ecological rationality to redressing the destructive relations between society and nature as they have evolved under Western modernity. For recent expositions on the Habermasian conception of communicative rationality and the implications for democratic practice, see T. F. Murphy, III, "Discourse Ethics: Moral Theory or Political Ethic," *New German Critique* 62 (1994): 111–137; and S. Chambers, "Discourse and Democratic Practices," in *The Cambridge Companion to Habermas*, ed. S. K. White (Cambridge: Cambridge University Press, 1995), 233–259.

3. Although the principal focus of this chapter is on the postwar period, we can find examples of the Frankfurt School's concern with ecological issues before the publication of *Dialectic of Enlightenment* (London: Verso, [1947] 1979). See the early essays of Max Horkheimer contained in the recently published collection (M. Horkheimer, *The Eclipse of Reason* [New York: Continuum, 1992]).

4. R. Eckersley, *Environmentalism and Political Theory* (London: UCL Press, 1992), 101–3.

5. D. Kellner, "Introduction," in *One Dimensional Man*, 2d ed., by H. Marcuse (London: Routledge, 1991), xxii.

6. H. Marcuse, *One Dimensional Man*, 2d ed. (London: Routledge, [1964] 1991).

7. H. Marcuse, "Ecology and the Critique of Modern Society," *Capitalism, Nature, Socialism* 3 ([1979] 1992): 29–48, quote 158, emphasis in original.

8. Some qualification is required over the use of the term positivism, which has a variety of potential meanings and applications. There are two main ways in which positivist doctrine has influenced environmental discourse: first, the promotion of a unified epistemological framework for nature and society based on the extension of the methodologies of the natural sciences; and second, the axiological tenet of scientific neutrality and value-freedom that pervades the environmental sciences. Jürgen Habermas and the intellectual legacy of critical theory have played a key role in developing the critique of positivism since the 1960s, but as I argue in this paper, his epistemological separation between nature and society is so abrupt as to limit its potential applicability for the advancement of environmental understanding. Though positivism has now been largely discredited as a coherent philosophical doctrine, it is still a useful term by which to highlight the misapplication of scientific method in the explanation of social phenomena.

9. W. Outhwaite, *Habermas: A Critical Introduction* (Stanford: Stanford University Press, 1994), 21-22.

10. J. M. Bernstein, *Recovering Ethical Life: Jürgen Habermas and the Future of Critical Theory* (London: Routledge, 1995), 29.

11. See J. Habermas, *The Theory of Communicative Action: vol.1, Reason and the Rationalization of Society* (London: Heinemann, [1981] 1984); J. Habermas, *The Theory of Communicative Action: vol. 2, The Critique of Functionalist Reason* (Cambridge: Polity Press, [1981] 1987); A. Honneth, and H. Joas, eds., *Communicative Action*, trans. J. Gaines and D.L. Jonas (Oxford: Polity Press, 1991).

12. See R. Kemp, "Planning, Public Hearings, and the Politics of Discourse," in *Critical Theory and Public Life*, ed. J. Forester (Cambridge, Mass.: MIT Press, 1985), 177-201.

13. Outhwaite, *Habermas*, 106.

14. J. Habermas, "Modernity versus Postmodernity," trans. S. Ben-Habib, *New German Critique* 22 (1981): 3-14.

15. See K. Soper, *What is Nature? Culture, Politics and the Non-Human* (Oxford: Blackwell, 1995).

16. U. Beck, *Ecological Politics in an Age of Risk* (Cambridge: Polity Press, [1988] 1995), 40.

17. A. Feenberg, *Alternative Modernity: The Technical Turn in Philosophy and Social Theory* (Berkeley: University of California Press, 1995), 222.

18. T. W. Luke, and S. K. White, "Critical Theory, the Informational Revolution, and an Ecological Path to Modernity," in *Critical Theory and Public Life*, ed. J. Forester (Cambridge, Mass.: MIT Press, 1985), 22-57.

19. J. Habermas, *The Philosophical Discourse of Modernity: Twelve Lectures* (Cambridge: Polity Press, [1985] 1987).

20. T. Hayward, *Ecological Thought: An Introduction* (Oxford: Polity, 1994).

21. On the extension of moral consideration in biocentric ethics, see R.

Attfield, "The Good of Trees," *Journal of Value Inquiry* 15, (1981): 35–54; A. Brennan, "The Moral Standing of Natural Objects," *Environmental Ethics* 6 (1984): 35–56; and P. Taylor, *Respect for Nature: A Theory of Environmental Ethics* (Princeton, N. J.: Princeton University Press, 1986). The environmental ethics literature has tended to be preoccupied with the impact of anthropocentrism on nature. More recently, however, a well-developed defense of the role of anthropocentrism in environmental ethics has emerged as part of a more nuanced critique of instrumentalist attitudes toward nature (see M. Sagoff, *The Economy of the Earth: Philosophy, Law, and the Environment* [Cambridge: Cambridge University Press, 1988]; B. Norton, *Towards Unity Among Environmentalists* [Oxford: Oxford University Press, 1991]; Soper, *What is Nature?*).

22. T. McCarthy, *The Critical Theory of Jürgen Habermas*. (Boston, Mass.: MIT Press, 1978); C. Alford, *Science and the Revenge of Nature: Marcuse and Habermas* (Tampa and Gainsville: Florida University Press, 1985).

23. S. Benhabib, and F. Dallmayr, eds., *The Communicative Ethics Controversy* (Cambridge, Mass.: MIT Press, 1990).

24. J. O'Neill, *Ecology, Policy and Politics: Human Well-Being and the Natural World* (London: Routledge, 1993).

25. B. Latour, *We Have Never Been Modern*, trans. Catherine Porter. (New York: Harvester Wheatsheaf, [1991] 1993), 60.

26. N. Smith, "The Spirit of Modernity and its Fate: Jürgen Habermas," *Radical Philosophy* 60 (1992): 23-29.

27. J. Habermas, *The Structural Transformation of the Public Sphere: An Inquiry into a Category of Bourgeois Society*, trans. T.Burger (Cambridge: Polity Press, [1962] 1989), 1.

28. J. Dryzek, "Ecology and Discursive Democracy: Beyond Liberal Capitalism and the Administrative State," *Capitalism, Nature, Socialism* 3 (1992): 18–42; see also J. Dryzek, *Rational Ecology: Environment and Political Economy* (New York: Blackwell, 1987).

29. Outhwaite, *Habermas*, 11. See also N. Fraser, "What's Critical about Critical Theory? Habermas and Gender," in *Feminism as Critique: Essays on the Politics of Gender in Late-Capitalist Societies*, ed. S. Ben-habib, and D. Cornell (Cambridge: Polity, 1987), 31–56; I. M. Young, *Justice and the Politics of Difference* (Princeton, N. J.: Princeton University Press, 1990); C. Calhoun, *Habermas and the Public Sphere* (Cambridge, Mass.: MIT Press, 1992); and D. Gregory, *Geographical Imaginations* (Oxford: Blackwell, 1993).

30. Smith, "The Spirit of Modernity," 28.

31. Eckersley, *Environmentalism and Political Theory*, 112.

32. Hayward, *Ecological Thought*, 46.

33. Beck, *Ecological Politics*, 2.

34. U. Beck, A. Giddens, and S. Lash, eds., *Reflexive Modernization: Politics, Tradition and Aesthetics in the Modern Social Order* (Oxford: Polity Press, 1994), 33.

35. U. Beck, *Risk Society: Towards a New Modernity* (London: Sage, [1986] 1992), 183, emphasis in original.
36. Beck, *Risk Society*, 186.
37. Beck, *Ecological Politics*, 8.
38. Beck, *Ecological Politics*, 182–83.
39. M. Rustin, "Incomplete Modernity: Ulrich Beck's *Risk Society*," *Radical Philosophy* 67 (1994): 3–11, quote 4.
40. Rustin, "Incomplete Modernity," 7.
41. U. Beck, *Ecological Enlightenment: Essays on the Politics of Risk Society*, trans. Mark Ritter (Atlantic Highlands, N. J. : Humanities Press, [1991] 1995), 14.
42. Beck et al., *Reflexive Modernization*.
43. Beck, *Ecological Politics*, 183.
44. See A. Feenberg, *Critical Theory of Technology* (Oxford: Oxford University Press, 1991); Feenberg, *Alternative Modernity;* and S. Helsel, "The Dialectic of Capitalist Technology," *New German Critique* 60 (1993): 161–169.
45. W. Leiss, *Under Technology's Thumb* (Montreal: McGill-Queen's University Press, 1990); 5.
46. Leiss, *Under Technology's Thumb*, 75; see also W. Leiss, *The Domination of Nature* (New York: George Braziller, 1972); and W. Leiss, *The Limits to Satisfaction: An Essay on the Problem of Needs and Commodities* (Montreal: McGill-Queen's University Press, 1988).
47. Beck, *Ecological Enlightement*, 60.
48. See P. H. Templet, and S. Farber, *The Complementarity Between Environmental and Economic Risk: An Empirical Analysis* (Baton Rouge: Louisiana State University Institute for Environmental Studies, 1992); US Environmental Protection Agency, *Toxic Release Inventory & Emissions Reductions 1987–1990 in the Lower Mississippi River Industrial Corridor* (Washington, D. C.: Environmental Protection Agency Office of Pollution Prevention and Toxics, 1993).
49. Beck, *Ecological Politics*, 56.
50. Beck, *Ecological Politics*, 7.
51. Beck, *Ecological Enlightenment*, 129.
52. Marcuse, "Ecology and the Critique of Modern Society"; Soper, *What is Nature?*
53. See D. Kellner, *The Persian Gulf TV War* (Boulder, Col.: Westview Press, 1992).

Critical Questions in Environmental Philosophy

Annie L. Booth

Aldo Leopold once wrote, "There are some who can live without wild things, and some who cannot."[1] Those who understand the truth in that statement are often compelled to seek solutions to the environmental problems that are leading us to that day when wild things will disappear. The field of environmental ethics is a possible source of those solutions. There is a two-thousand-year history of environmental law and policy, and two thousand years of partial success.[2] It became clear to many that an essential component was missing: an understanding of how people, at the individual and cultural level, understood the natural world and their relationships with it. This is the premise at the center of ecological philosophy.

That the human-natural world intersection is vital has for many become a given. So, too, is the contention that the natural environment and human communities are facing potentially fatal disturbances. That we need to change the ways we go about our lives is also a recognition no longer confined to obscure journals. Thus, the field of environmental ethics, or environmental philosophy, has experienced a huge growth in interest in recent years, not just in academic circles but among a concerned public and increasingly among government agencies and industries.

In the rush to consider new ways of problem solving, it is tempting to overlook critical problems developing in the ideas under consideration. This article examines some of the key problems developing in two related but distinct branches of environmental ethics: bioregionalism and ecological feminism. Such a discussion is not meant to discredit environmental ethics as a solution. Indeed, the ideas contained within bioregionalism and ecological feminism offer exciting opportunities to reconsider not only human ethics, but environmental policy and regulation as well as natural resource management. Instead, by illuminating certain problems, I hope to suggest ways in which environmental ethics can be made more relevant and usable to policymakers and resource managers.

In addition, for at least one of the two environmental philosophies under discussion, bioregionalism, the example and precedent of indigenous peoples have been of considerable significance in formulating theory. The "environmental movement" as a whole has been interested in the example of Native Americans from its earliest antecedents and remains caught in a fascinated if problematic relationship with indigenous cultures.[3] Although bioregionalists employ indigenous perspectives as examples of appropriate relationships between humans and the natural environment, that argument itself requires critical examination. This paper will touch briefly on some of the issues that need to be scrutinized before indigenous perspectives are imported wholesale into environmental philosophy.

To place such a critical discussion in context, however, a brief overview of the shared central nature of varieties of environmental ethics as well as some specifics of bioregionalism and ecological feminism are warranted.

What Is Environmental Ethics?

Environmental ethics, or environmental philosophy, can be described as an exploration of the cosmos and humanity's relationship to it. Where environmental ethics could be said to differ from other explorations is in its explicit focus on the natural world. Environmental philosophy is the marriage of ecology and philosophy.

Environmental ethicists build on ideas that go as far back as the pre-Socratic philosophers. What links the ideas and theorists as environmental philosophers is an effort to construct a theory of humanity that is located within the natural world and that describes human-nature interactions as something other than exploitative, hierarchical, or resource-oriented. Early efforts included the works of Henry David Thoreau, John Muir, Aldo Leopold, and Rachel Carson. However, environmental ethics as a formal field of study has its beginnings in the

1960s environmental movement and the widespread public acknowledgment of ecological problems. This acknowledgment led to such widely publicized events as the first Earth Day in 1970, and the United Nations Conference on the Environment, held in Stockholm in 1972. Somewhere between those two events, environmental philosophy appeared as a legitimate field for intellectual exploration. The first forum was a conference combining philosophy and environmentalism held at the University of Georgia in 1971. The proceedings of that conference were later edited into one of the first recognizable academic books in environmental philosophy: *Philosophy and Environmental Crisis*.[4]

This early work established one significant direction of the larger debate in environmental philosophy: whether nature could be considered a moral agent, or, in other words, whether nature was a thing to which humans might owe obligations or have duties toward. Discussion revolved around such issues as animal rights, the extension of legal rights to natural objects, intrinsic versus instrumental values of nature, and the question of stewardship over nature. The early 1970s also witnessed the emergence of humanistic environmentalism and psychological environmentalism. These issues continue to represent a significant portion of the literature in environmental philosophy today.

During the early 1970s, some traditions within environmental philosophy also developed, including bioregionalism (1972) and ecological feminism (1974). During the same period there was a revival of interest in Native American traditions signaled by the 1973 publication of a key article, Stewart Udall's "Indians: First Americans, First Ecologists," and reinforced by the popularity of certain books on Native American philosophy.[5] Several environmental philosophers took Native Americans to be an example of appropriate human-nature relationships.

While the 1980s and 1990s have witnessed the continued growth of environmental philosophy, its focus has narrowed with time and changing interest. There is a growing body of literature devoted to Christian-based environmentalism. Ecological historicism has emerged. Debates continue over animal rights and the intrinsic value of nature. I will focus in this paper on the major debates centering on bioregionalism and ecofeminism.

Two Varieties of Environmental Ethics

Bioregionalism

Although its roots go back much further, bioregionalism in its present form has been around since 1972. Inspired by the 1972 U.N.

Conference on the Environment, Peter Berg was the first to try to bring together people with shared ideas. Berg drew upon the term, "bioregion," described first by expatriate American Alan Van Newkirk, who argued that to undertake effective conservation, one had to work with the boundary lines drawn by nature rather than politicians. "Natural resources" and ecological problems ignored national and political boundaries. The term "bioregion" reflected the more appropriate focus on natural boundaries.

Bioregionalism grew out of a number of previously existing ideas. It has roots in the theories of regional planning going back to the end of World War II. Kirkpatrick Sale, a prominent advocate of bioregionalism, cites the influence of planners and thinkers such as Lewis Mumford, Jane Jacobs, Leopold Kohr, Ian McHarg, and E. F. Schumacher.[6] The influence of biogeographers can also be noted in bioregional theory, and such transborder management efforts as the Migratory Bird Treaty (1916) or the Convention for the Protection of Birds Useful to Agriculture (1902).[7] Other, older roots are the worldviews and philosophies of Native American tribes. Bioregionalists are open in their admiration for societies who managed to live, as bioregionalists believe, in harmony with the land. Part of the focus of bioregional research involves the explication of how resident tribes lived. Bioregionalism is a deliberate attempt to create in all Americans the sense of being "indigenous" to a place.

Bioregionalism centers around the identification of, and with, a bioregion. A bioregion is a piece of land that is defined by physical, biological, and cultural characteristics. As Berg and Dasmann note, "The term [bioregion] refers both to geographical terrain and a terrain of consciousness—to a place and the ideas that have developed about how to live in that place."[8] Bioregionalism implies more than merely identifying units of distinctive cultural-biological characteristics, however. It means discovering and living within the constraints and possibilities imposed by the character of the bioregion. It means living in a manner that is, to use a modern buzzword, "sustainable" over time. Thus, the bioregionalist does not attempt to grow citrus fruits and grass lawns in the desert, but looks instead at what *can* grow naturally in the region, what native resources can be developed, and what lifestyles can be sustained within the dictates of the region's characteristics.

Bioregionalists argue that community, human and natural, is the level at which living should be done. Implicit in bioregionalism is the idea of decentralization, in both political and economic spheres. People who live-in-place are those who, logically, have the best sense of what is right for that region, and who, therefore, should make the decisions affecting the region. However, an interest and devotion to the community and region do not imply disinterest in the world outside. This complex mix of philosophy and practicality, poetry and community, is the heart of bioregionalism as it exists today.

Ecological Feminism

The term "ecofeminism" was coined by French feminist Françoise d'Eaubonne in 1974 in an effort to encourage women to believe in their ability to bring about an ecological revolution.[9] The term has since been expanded far beyond what d'Eaubonne probably envisioned. Ecofeminism draws upon efforts during the 1970s to develop a "feminist" consciousness, hence the focus on "liberating" women from their oppression. Ecofeminism also reflects a concomitant interest in "feminist" spirituality, which developed at the same time, and linkages with the environmental movement of the early 1970s.

The main threads of ecofeminism developed during the late 1970s and early 1980s. Key proponents to pick up on d'Eaubonne's term include Susan Griffith (poet and author), Ynestra King (activist), Elizabeth Dodson Gray (theologian), Hazel Henderson (economist), Carolyn Merchant (academic), and Rosemary Radford Ruether (theologian). Karen Warren, Charlene Spretnak, Jim Cheney, and Michael Zimmerman arrived in the field in the late 1980s, as ecofeminism became a legitimate topic for discussion within the academe.

There is no one single teaching or belief that characterizes ecofeminism, but instead there are several interrelated beliefs from which theory is developing:

1. Women have different ways of seeing and relating to the world around them which offer unique insights on interactions between humans and the natural world.
2. These female insights (which are tendencies rather than absolutes) have been ignored, devalued, or suppressed in cultures and societies dominated by males and their values (patriarchies), as women have been devalued.
3. The suppression and domination of women and the domination and control of the natural world are connected. Women are controlled because they are thought to be closer to primitive

nature, and the control of wild nature is justified by its personification as female.
4. By understanding this connection and by revaluing and exploring feminine ways of seeing and relating, both women *and* men can discover positive ways of interacting with the natural world and with each other.[10]

Ecofeminists (and feminists) work toward dismantling a worldview they label "patriarchy." When ecofeminists talk about patriarchy, they are talking about a culture or society in which men, and male experiences and activities, are paramount. If the experiences or activities are part of female life, only those experiences will automatically be discounted as less valuable and important. In a patriarchy, women automatically hold lesser positions in society, in politics, and in cultural events. Patriarchy is, essentially, a way of seeing the world and others in which "good" is substituted for "male" and "bad" for "female." Male experience is projected outward onto society and, as Henderson says, it is "universalized . . . as if it were *human* experience. Of course it is not. . . ."[11]

A second key debate in some ecofeminist circles comes out of the ecofeminist critique of patriarchy. There is crucial debate over whether or not men and women have qualitatively different perceptions. Is the world a different place for those of different biological gender? Or are these differences culturally created by patriarchal oppression of women and others? This is a thorny, divisive issue for ecofeminists and one that is dangerous as well, as it holds the potential of returning women to a prefeminist "natural woman" sensibility. One camp of ecofeminists, including Biehl and Griscom, argue that there are no real differences. Men and women perceive in equivalent ways and neither has better claim to relating to nature. Any apparent differences are socially conditioned.[12] Another camp, which includes Hallen, Spretnak, Gray, and Henderson, argues that there are distinct biologically based differences. These differences are the basis for much ecofeminist theorizing, which suggests that women have the potential for a more positive relation with nature than do men.[13]

Flaws in Bioregional Theory

Bioregionalism is, as one geographer notes, "a moral philosophy, sometimes romanticized as a 'system of thought,' a framework for action that celebrates geographic and cultural diversity, the sacredness of the Earth, and the responsibilities of local communities to it."[14] The danger lies in romanticizing. For all its practical bent, bioregionalist theory has yet to come seriously to grips with important questions and problems, including the crucial questions of population levels and distribution.

In most countries, populations are too high and the land base, even with equitable redistribution, too small to permit everyone to go back to the land, even if everyone wanted to. Nor could people go back to the land without substantially modifying their ecosystems. In other words, in most countries, bioregionalist theories are not going to permit the salvation of much wild nature, or species that will compete in some way with human needs. Elephants and tigers (or grizzlies and cougars) are not now or in the future going to be welcome residents in human areas. Bioregionalists counter by arguing that decentralization would only go so far, and that most humans would dwell in intensely concentrated communities of perhaps five thousand to ten thousand.

In fact, a new strand of bioregionalism specifically devotes itself to designing "greener" cities.[15] These communities would then be surrounded with areas of agriculture and wild nature. However, the question remains whether the small nations, let alone the United States, even if governed by judicious regionally centered interests, have the land base to support such a population redistribution. Perhaps, if everyone were sufficiently willing, and if all areas were created equal.

Bioregionalism recognizes the necessary limitations physical conditions place on development activities. However, it is obvious that not all areas will support similar population concentrations. In the desert states, especially around Las Vegas and Los Angeles, the population is now supported only by massive infusions of water. A bioregionally self-sustaining population in this region would be forced to shrink. How will the inevitable redistribution of population occur? Will volunteers be requested? Will only the longest term residents be permitted to stay? Who will make the decision? And where will all those people in Los Angeles and Phoenix and Las Vegas go? To other bioregions that may already have more people than *they* can support?

Further, many areas are lacking significant reserves of essentials for living, not by natural forces but as a result of market forces. Hundreds of years of colonialism and neocolonialism have witnessed any number of weaker political entities stripped of natural resources to supply the stronger countries. How will these predominantly Third World, previously stripped, areas respond if we of the First World now suggest they live in a self-sufficient manner? Reinhabitation does not allow for, even if it were possible, the return of plundered wealth.

Bioregionalists are often accused of being politically naive. Coleman argues that any unilateral movement toward regionalism only opens up the possibility that another megalomaniac with dreams of empire will come along and gobble the region up: "In fighting giantism we need a strategy. Regionalism can easily turn out to be the decentralist's Trojan horse. Supergovernments, by breaking up national governments into small pieces, can pick off the pieces one by one, leaving all the real decision-making at the continental bloc level [much like the North American Free Trade Agreement]."[16]

The similarity of this argument to those against nuclear disarmament, particularly unilateral disarmament, is probably not an accident; both fears stem from a common root: us against them. In this, bioregionalists are probably most distinct from the most radical separatist movements, in that bioregionalists seem to feel the need to retain some umbrella nation-state structure while focusing most powers at the regional level. How this political setup would evolve and function has not been clearly set out, although it would require revolutionary changes in political structures.

Bioregionalists often operate with the assumption that people are inherently good. Bioregionalists therefore often gloss over the question of what kind of people are going to be running things in these bioregions. Those with less certainty about people might argue that central governments are essential to ensure someone somewhere is not unfairly discriminated against or worse as a result of gender, sexual preference, race, ethnicity, age, and so on. Sale, however, disagrees:

> I've even heard people challenge bioregionalism by saying: You'll always have to have a strong central government to be sure that racists, or homophobes, or sexists, or polluters (or whatever bugbear is their fashion) don't come to power in any given bioregion—as if such people weren't in power now, as if an empowered bioregion would automatically advance such people, as if the central government had proved itself such a wonderful enemy of racism and sexism and pollution that we ought to keep it around.[17]

Perhaps. But as Salibian notes (and anyone who has resided in a small town might confirm) small enclaves and units of authority often are more agonizingly tyrannical than anything a big government could impose.[18]

Further, weaknesses in centralized governments do not necessarily justify a belief that smaller units will be better. While no one is suggesting that we return to insular uninformed settlements (although this is a trend already under way), and while it is assumed that bioregionalists will pay considerable attention to the world perspective while working on their regional concerns, bioregionalists need to be more aware of historical precedent. *Small* feudal monarchies, in which neither human nor ecological needs or rights were recognized, have always existed.[19] Questions of political authority and rights are issues to which bioregionalists must pay more attention.

Perhaps these issues are not addressed sufficiently as bioregionalists themselves are struggling with practical questions of adequately addressing the needs of people other than white middle-class Americans. Bioregionalism has been criticized for being indifferent to questions of poverty and racial inequity. By the Third Bioregional Congress in 1988, the bioregional movement was making an explicit effort to incorporate such concerns, but this effort, rather like the effort for gender equity, has come very late in the process.[20] The bioregionalists appear to be somewhat threatened by losing control over "their" movement. They often celebrate in the abstract the contributions of people of color, for example, Native Americans. However, they have failed to reach out to, and acknowledge, the large numbers of communities that have existed "bioregionally" for generations. Rather, there has been a desire to offer bioregionalism as a solution to poor people who are already practicing the ideas of bioregionalism under another name. An African-American participant at the Third Congress described it this way:

> I would encourage folks, particularly those already involved in the movement, if you are interested in really broadening the base of bioregionalism, then it is essential for white people to really get to know people of color and to do that without an agenda. That white people need to listen in order to learn and must risk being changed by people of color in their communities. That becomes a foundation for some sort of mutuality and the ability to work together.[21]

Last, bioregionalists often ignore the fact that many small "natural" political entities originally existed because of strong ties of ethnicity, religion, and language. Such entities are what bioregionalists are trying to recreate, but they miss one quite logical consequence. For all the good of community ties built on such a basis, there is an equal likelihood that such ties will lead to significant hostility toward outsiders. At worst, it becomes a cause for war. Much of the strife in the Commonwealth of Independent States is based on long-standing ethnic and religious differences, as is the conflict between the Serbs, Bosnians, and Croatians, Northern Ireland and Ireland, Israelis and Arabs, or Rwanda, differences that bioregionalists wish to celebrate. This lapse in attention to political realities remains a critical gap in bioregional theory. How does one solve a problem that is global in scope and incredibly ancient? As one critical commentator stated, "It's an old, old dream that a return to nature will simplify and clarify human affairs . . . To make geography the basis of society is to oversimplify the complexity of human nature."[22]

In spite of all its drawbacks and undeniably idealistic bent, bioregionalism is a positive direction in environmental ethics. It offers creative alternative visions of what life could be like. It is these visions, many of which *are* achievable now, which are perhaps most valuable. It is no longer a question of thinking "what if," but "why not?" To ask that question is a valuable contribution indeed. Bioregionalism is valuable as well because it is accessible, understandable, and appealing to almost everyone. Many people dream of less-congested, greener cities; of going out to the country and living more simply; of feeling as if you belong to a place, rather than feeling lost and alienated; and of having the power to make a difference. It is that vision the bioregionalism speaks to effectively.

Finally, bioregionalism is a *practice*, perhaps in search of a theory, but a practice nevertheless. When asked how to change the world, bioregionalists do not quote philosophers or discuss political theory. Instead, they point to concrete examples, examples that they have had a hand in creating: alternative communities, sustainable farms, human-based, ecologically minded dwellings suited to land conditions. And that practicality is perhaps the most appealing quality of all.

Ecofeminist Difficulties

Ecofeminism is such an internally riven, convoluted set of ideas that it is difficult to point out critical gaps, since a problem for one ecofeminist theorist will be repudiated and readdressed by another. But perhaps this is its largest flaw: so many theories and theorists label themselves "ecofeminist" that it becomes difficult to describe what ecofeminism is. A tolerance for diversity is a good thing, in general, in contrast to one hard and fast dogma, but as Janet Biehl once commented, "dogmatism is clearly not the same thing as coherence, clarity, and at least a minimum level of consistency."[23]

One of the greatest criticisms of ecofeminism, and one that will not be resolved any time soon, is the tendency to argue dualisms. *Either* female experience and nature is a biologically based reality *or* it is a social construction. *Either* we focus on the construction of social and political alternatives *or* we work on spiritual issues, including personal redemption. *Either* men are wholly responsible for any number of crises, including the ecological crisis, *or* . . . well there is no "or" here. Men in technoindustrial patriarchy are, with some exceptions, simply the "bad guys"; ecofeminists have made few efforts to find any redeeming quality in men as a group. In fact, this is a key criticism of ecofeminist theory; it has so little place or consideration for men. Grounding politics or spirituality only in the (possibly) unique experiences of one gender is no less reactionary or bigoted because that experience is female than it is when the only validating experience is male. Ecofeminists run the risk of becoming as hierarchical as the patriarchy they seek to overthrow.[24]

There is no dualism either in the fact that most ecofeminists are white and middle-class, and it has only been quite recently that efforts have been made to include people of color or to engage with their issues. Thus, ecofeminists often talk of "women's" issues or "women's" reality with little recognition that issues and reality for a white, economically comfortable, North American academic might be very different from issues of concern to a woman of color in inner-city Chicago or a woman in famine-ridden Ethiopia. This insularity is changing, but slowly. As King and Warren have recognized, "feminism must be a 'solidarity movement' based on shared beliefs and interests rather than a 'unity in sameness' movement based on shared experiences and shared victimization."[25] However, ecofeminism has spent much of its time arguing that all women are oppressed by patriarchy, as is nature, and both are exploited for the same reasons: in other words, ecofeminists' focus has been on defining similarities between women and their socioeconomic conditions. But the reverse, stressing the difference between different women, should not result in violently oppositional dualisms, as it often seems to in ecofeminism.

The dualism between spiritual ecofeminists and political or social ecofeminists is particularly troubling, as there is nothing inherently oppositional between the two groups. Even in the "patriarchal" religions there is a strong tradition of social and political activism that had a grounding in spiritual beliefs: Christ, Gandhi, Martin Luther King, Jeanne d'Arc, the Christian families and churches that hid Jews during the Holocaust—there are uncountable examples. But at the same time, what is social life without some sort of sense of a broader set of ideas and values? It does not have to be spirituality that provides this, but it often is. Rationality and logic take us so far; eventually there comes a need for a great emotional leap of intuition. Spirituality often provides the lift for just such a leap.

True, social and political activism and religion do not always lead to positive results. However, that is not necessarily a reason for declaring one or the other as dangerous or useless. To insist on rigid distinctions, as some ecofeminists seem inclined to do, is to believe that a black and white picture is an appropriate coloring of the world. But most often, the decisions and the issues are painted in shades of gray. Why *not* the responses?

The bitter, personal exchanges that this dualism has occasioned are equally problematic. Agreeing to disagree while working together and respecting each other's opinions seems to be beyond some ecofeminists, let alone between ecofeminists and others, such as deep ecologists. And yet, these are people who, supposedly, should have a commitment to cooperation. They have after all a shared goal: a socially and ecologically sound world. Unfortunately, for all their discussion of cooperation, the importance of relationships and loving care, ecofeminists seem to be as willing as the next patriarch to treat others as objects with nothing of value to contribute. There is a lingering and regrettable flavor of turf wars, one-upping the other critic, and self-righteousness in much ecofeminist writing. This wrangling points to two difficulties. First is the difficulty, and not only among ecofeminists, of making a transition between theory and action. Humanity has always confronted the dilemma of facing contradictory needs and outside pressures. Second is the contradiction of attempting to make change while forced to operate in a system that functions in a completely different way. Women's studies departments in universities are often good examples of this problem. Women's studies departments spend a great deal of time teaching and writing about oppression, techniques of domination, and the need to change to more cooperative ways of doing things. At the same time, the faculty within these departments are as capable as any other academics of tyrannical, oppressive behavior and can be as engaged in academic turf battles. Of course, they operate in a system where tenure and publication requirements encourage this sort of behavior. That they succumb to the system is indicative of some of the enormous obstacles that the ideals of ecofeminism face when confronted by everyday reality.

Finally, ecofeminists often do themselves a great disservice in promoting and developing their ideas by their rush toward legitimacy. To make a point, many ecofeminists seem too willing to play fast, loose, and careless with history, with science, and with religion. They are often given to categorical and critical statements that present only a limited side of things. For example, some ecofeminists have tried to demonstrate legitimate, historically based grounds for the idea of women who had high status and were fully functional social and political beings. It is an understandable goal but to fulfill it, ecofeminists such as Eisler and Spretnak have relied on data that are not as incontrovertible as they have presented it. Arguing that women's domination began with the fall of the Neolithic women-centered cultures by using only one contested source of information leaves writers open to criticism that many are only too happy to give.[26]

Further, it is important that ecofeminists avoid the trap of seeking so hard to legitimize themselves that they come to believe their ideas are best expressed in obscure academic journals in language that requires at minimum a doctorate in philosophy. Ecofeminism, along with other ecophilosophies, will only succeed when it reaches and captivates as many people as possible, especially those outside of the academe. Otherwise, the people most in need of finding new ways of thinking, or new words that express what they have already felt but couldn't communicate, are going to dismiss ecofeminism. And that would be everyone's loss.

For all its flaws and definitional slipperiness, ecofeminism makes a positive contribution to environmental philosophy. Its practitioners were the first to raise the question of women's social status as an important consideration in the environmental movement. Their pressure put women's issues and ideas into the agendas of other environmental philosophies. Ecofeminists raised questions that made both women and men take a hard look at the way they operated and at means as well as ends. Most important, it gave back to 52 percent of the human race a sense that they can contribute to the search for a better world without having to play the unsuccessful game already in place.

A Problematic Inspiration: Native American Worldviews

As mentioned earlier, environmentalists and environmental philosophers, especially bioregionalists, have long been interested in Native

Americans. The examples of "appropriate lifestyles" and "nature friendly" cultures that indigenous peoples offer are often held up by environmental philosophers as directions that technoindustrial cultures should perhaps try to follow. The interest in such an "indigenous" ideal raises a great number of questions that few environmental philosophers appear to be willing to address. I will only raise a few of these questions, but I suggest that the environmental philosophers must come to understand the difference between intellectual plundering, intentional or otherwise, and respectful comparative study.

One key issue that must be understood is the reluctance that indigenous peoples themselves feel toward becoming an "example." There have been too many misunderstandings and problems for Native Americans, of whatever culture, to say they even halfheartedly support environmentalists. As Ojibwa Winona LaDuke suggests, for too long environmentalists had difficulty in understanding the role Native Americans could play in environmental issues: "Generally, the environmental movement . . . doesn't really understand the relationship indigenous people have to the environment, and as a result has a tendency to separate the two, to view indigenous people as not having a right to our natural resources or not having a right of our lands in the interest of preserving the environment."[27] Even today, there is little representation of native groups in either the directing board, staff, or membership of environmental groups. For example, The Nature Conservancy, a well-funded land acquisition group, has one Native American staff member who is responsible for the liaison with all Native American groups in the United States.[28]

While the environmental movement has been slow to work with many Native American groups or issues, it has not hesitated to use their philosophy insofar as it is understood. The influence is clearly present in bioregionalism, as well as in more mainstream groups such as the Sierra Club. In all cases, Native Americans have been taken to be a model of "appropriate" lifestyles, practices, or spiritual beliefs, appropriate because they are seen as environmentally benign and because their beliefs are somehow "better," more "authentic," than the corrupt technoindustrial beliefs ecofeminists and bioregionalists are attempting to dismantle.[29] But this sense of appropriateness is problematic.

In the first place, such claims deny or ignore the very real and difficult questions facing the tribes today, including the need to make a living, sometimes through the exploitation of natural resources. It assumes there is one basic "good" way of acting, and that given the chance, all Indians will go back to being "natives." Environmentalists often do little to restore damaged lands, or to alleviate massive unemployment on reservations, or to provide adequate health care to people whose standard of living is roughly equivalent to that in many poor Third World countries. Using Native Americans as a model condemns them to remain faceless images with no real voice in the present or the future.

Second, most environmental philosophers are making a decision on the propriety of native lifestyles with no real knowledge if what they are claiming is appropriate. They make passing references to the examples of "primal peoples" and perhaps the use of a couple of sources to describe hundreds of different cultures.[30] It is an utterly incomplete set of data from which environmental philosophers are drawing their conclusions. Further, there is an almost insulting romanticism in the view that many environmental philosophers have of indigenous peoples as "gentle people living in harmony with nature." The reality was far more complex and problematic than that. Perhaps it is overstating the case, but often environmental philosophers (and environmentalists) seem more comfortable with the long-dead Indian, than the more prickly and less cooperative (or cooptable) living item.

Third, there is a question of how well the examples such cultures offer us would translate into predominantly white, technoindustrial cultures. Brown suggests that seeking enlightenment from an alien tradition poses significant problems. Too often we are unwilling to approach a tradition on its own level, within its own structures and languages. What we "find" in an alien tradition is a projection of what we find missing in our own traditions. Individuals in a particular tradition spend years preparing to participate. They prepare through their life-long participation in language, which confers sacred power through its vocabulary, and through structures and categories of thought that serve as vehicles of transmission for traditions that are only practiced over time and with sacrifice. The tradition is lived on a day to day basis.[31] As Toelken remarks, using the Navaho as his example:

> The system is impressive to us because of its assumption that all phenomena are integrated and interdependent, not extractable and abstractable from each other. . . . Rather than viewing the Navajos as some kind of primitive desert "flower children," moving easily through a harmonious world because of their recognition of relationship among all things in nature, we need to recognize that the Navajos are participants in an

extremely rigorous philosophical and ritual system which places demands on individuals that most non-Navajos would find it difficult to cope with.[32]

Yet people today are seeking wisdom from native traditions in greater numbers than ever before. In our efforts to find an answer, any answer, that will fulfill our ecological and personal needs, we have become involved in cultural exploitation, much as our ancestors exploited them for their land. We use Native American beliefs as if the beliefs were free for the taking, and as if the owners no longer had any use or any rights to them. They have become commodities to be extracted and sold as the economic demand appears. LaDuke suggests, "We gotta learn to relate to the earth without the stealing of indigenous culture":

> I'm not sure why in order to apprehend the fact that your life is contingent upon the life and the environment around you, why it entails that you go and pretend to be a Lakota in a sweat lodge, or that you purport to teach your neighbors Navajo crystal healing ceremonies. I just don't see the relation between the two. So to me the answer is very simple. It's done because you want to pretend you're something that you're not rather than becoming something else.[33]

That is a challenge that many in the environmental philosophy community, and in the environmental community as a whole, have been slow to face.

Some Conclusions?

Definitive conclusions on the current state or the possible futures of environmental philosophy are difficult to draw. Environmental philosophy is a little over two decades old. The theories of ecofeminism and bioregionalism are in a constant state of flux as they continue to evolve. Their final constructs will be interesting, but possibly long in emerging, if indeed they ever emerge. Perhaps the importance of these emerging worldviews is not in their end result but in the process they generate. Asking questions of ourselves, of our beliefs about the natural world and our relationship to it, is the process that environmental philosophy encourages. Whether we agree with any one model or parts of a model may be irrelevant if we are willing to be drawn into the process itself.

I have criticized ecological feminism and bioregionalism, not because I disagree with their proponents' intentions, but because I believe environmental philosophers need to take some time from continual expansion of their theories and look back and evaluate what already exists. It sometimes appears that many environmental philosophers forget one of the important lessons that the study of ecology must teach us: the importance of cooperation and complexity. At several points, both environmental philosophies succumb to isolationism or Darwinian competitiveness, "my theories are the correct ones," or occasionally to monoculture, "all theories are part of my theory." I submit that just as ecosystems appear to thrive through symbiosis, diversity, and, often, cooperation, so too might the field of environmental philosophy.

Notes

1. Aldo Leopold, *A Sand County Almanac* (New York: Ballantine Books, [1949], 1981), xvii.
2. See Paul S. Martin, "Prehistoric Overkill," in *Man's Impact on the Environment,* ed. Thomas R. Detwyler (New York: McGraw-Hill, Inc., 1971), 612–24; N. C. Pollock, *Animals, Environment and Man in Africa* (London: Saxon Books, 1974); J. Donald Hughes, "Mencius' Prescriptions for Ancient Chinese Environmental Problems," *Environmental Review* 13, no. 3–4 (fall/winter 1989): 15–28; J. Donald Hughes, *Ecology in Ancient Civilizations* (Albuquerque: University of New Mexico Press, 1975); John Livingston, *One Cosmic Instant* (New York: Dell Publishing, 1973); and Don Bronkema, "It Didn't Happen Yesterday," *EPA Journal* 14, no. 4 (May 1988): 33–34.
3. See Stewart L. Udall, "Indians: First Americans, First Ecologists," in *Readings in American History—1973/1974* (Conn.: Dushkin Publishing Group, 1973); Doug Boyd, *Rolling Thunder* (New York: Dell Publishing, 1974); John (Fire) Lame Deer and Richard Erdoes, *Lame Deer: Seeker of Visions* (New York: Simon and Schuster, 1972); and John G. Neihardt, *Black Elk Speaks: Being the Life Story of a Holy Man of the Oglala Sioux* (New York: Pocket Books, [1932] 1975). For a critique of this, see Winona LaDuke, "Environmentalism, Racism, and the New Age Movement: The Expropriation of Indigenous Cultures," text of a speech, *Left Green Notes* 4 (September/October 1990): 15–18, 32–34.
4. William T. Blackstone, ed., *Philosophy and Environmental Crisis* (Athens: University of Georgia Press, 1974).
5. Udall, "Indians: First Americans, First Ecologists."
6. Kirkpatrick Sale, *Dwellers in the Land: The Bioregional Vision* (San Francisco: Sierra Club Books, 1985).
7. See Alfred Kroeber, *Cultural and Natural Areas of Native North*

America (Berkeley: University of California, 1939); or Carl O. Sauer, *Man in Nature: America Before the Age of the White* (Berkeley: Turtle Island Foundation, 1975).

8. Peter Berg and Raymond F. Dasmann, "Reinhabiting California," in *Reinhabiting a Separate Country*, ed. Peter Berg (San Francisco: Planet Drum Foundation, 1978), 217–20.

9. Françoise d'Eaubonne, *Le Feminisme ou la Mort* (Paris: Pierre Horay, 1974), 204.

10. These principles are derived from Ynestra King, "Towards an Ecological Feminism and a Feminist Ecology," in *Machina Ex Dea*, ed. Joan Rothschild (New York: Pergamon Press, 1983), 118–129; Ynestra King, "Ecofeminism: On the Necessity of History and Mystery," *Women of Power* 9 (spring 1988): 42–44; Charlene Spretnak, "Ecofeminism: Our Roots and Flowering," *Women of Power* 9 (spring 1988): 6–10; and Karen J. Warren, "Feminism and Ecology: Making Connections," *Environmental Ethics* 9, no. 1 (1987): 3–20.

11. Hazel Henderson, "The Warp and Weft: The Coming Synthesis of Eco-philosophy and Eco-feminism," in *Reclaim the Earth,* ed. Leonie Caldecott and Stephanie Leland (London: The Women's Press, 1983), 203–214, emphasis in original.

12. Janet Biehl, *Finding Our Way: Rethinking Ecofeminist Politics* (Montreal: Black Rose Books, 1991); Joan L. Griscom, "On Healing the Nature/History Split in Feminist Thought," *Heresies # 13* 4, no. 1 (1988): 4–9.

13. Patsy Hallen, "Making Peace With Nature: Why Ecology Needs Feminism," *The Trumpeter* 4, no. 3 (1987): 3–14; and Charlene Spretnak, "Introduction," in *The Politics of Women's Spirituality: Essays on the Rise of Spiritual Power Within the Feminist Movement,* ed. Charlene Spretnak (Garden City: Doubleday 1982), xi–xxx.

14. James J. Parsons, "On 'Bioregionalism' and 'Watershed Consciousness,' " *The Professional Geographer* 37, no. 1 (1985): 1–6.

15. See Peter Berg, Beryl Magilavy, and Seth Zuckerman, *A Green City Program for San Francisco Bay Area Cities and Towns* (San Francisco: Planet Drum Books, 1989).

16. John Coleman, "The Common Market—Another Superpower?" in *How To Save The World: A Fourth World Guide to the Politics of Scale,* ed. Nicholas Albery and Mark Kinzley (San Bernadino: The Borgo Press, 1984), 70–74.

17. Kirkpatrick Sale, "Devolution American-Style: Decentralization Hits Home," *Utne Reader* 30 (November-December 1988): 95–97.

18. Sunny Salibian, "Gently Persist," in *How To Save The World*, 159–62.

19. Donald Alexander, "Bioregionalism: Science or Sensibility?" *Environmental Ethics* 12, no. 2 (1990): 161–73.

20. The first congress was held in 1984 and the second in 1986.

21. Paul Cienfuegos, "On the Authentic Inclusion of People of Color—An Interview with Yanique Joseph and Jeffrey Lewis," *Third North American Bioregional Congress Proceedings*. 21–26 August, 1988, British Columbia (San Francisco: North American Bioregional Congress, 1988), 24–27.

22. Rosalind Williams, "Earth Mother Knows Best: A Review of 'Dwellers in the Land.' " *New York Times Book Review* (October 1985): 15.

23. Biehl, *Finding Our Way*, 3.

24. Carolyn Merchant, "Ecofeminism and Feminist Theory," in *Reweaving the World: The Emergence of Ecofeminism,* ed. Irene Diamond and Gloria Feman Orenstein (San Francisco: Sierra Club Books, 1990), 100–105.

25. Karen J. Warren, "The Power and the Promise of Ecological Feminism," *Environmental Ethics* 12, no. 2 (1990): 125–46; Ynestra King, "Healing the Wounds: Feminism, Ecology, and the Nature/Culture Dualism," in *Reweaving the World*, 106–121.

26. Riane Eisler, *The Chalice and the Blade—Our History, Our Future,* (San Francisco: Harper and Row, 1987).

27. LaDuke, "Environmentalism," 18.

28. Donna House, Native Affairs Coordinator, The Nature Conservancy, Private Communication.

29. See, for example, Cate Sandilands, "Ecofeminism and Its Discontents: Notes toward a Politics of Diversity," *The Trumpeter* 8, no. 2 (Spring 1991): 90–96.

30. There are three popular references. The first is a speech by Duwarmish Chief Seattle, which appears to have been something of a fabrication. See Rudolf Kaiser, "A Fifth Gospel Almost: Chief Seattle's Speech(es)," in *Indians and Europe,* ed. Christian Feest (Aachen: Rader Verlag, 1985), 505–526. Another is T. C. McLuhan, *Touch the Earth: A Self-Portrait of Indian Existence* (New York: Outerbridge & Lazard, Inc., 1971). The last is Luther Standing Bear, *Land of the Spotted Eagle* (Lincoln: University of Nebraska Press, 1933). For further references see Annie L. Booth and Harvey M. Jacobs, *Environmental Consciousness—Native American Worldviews and Sustainable Natural Resource Management: An Annotated Bibliography*, no. 214 (Chicago: Council of Planning Librarians, 1988).

31. Joseph Epes Brown, *The Spiritual Legacy of the American Indian* (New York: Crossroad Publishing Co., 1985).

32. Barre Toelken, "The Demands of Harmony," in *I Become Part Of It: Sacred Dimensions in Native American Life,* ed. D.M. Dooling and Paul Jordan-Smith (New York: Parabola Books, 1989). 59- 71.

33. LaDuke, "Environmentalism," 33.

Index of Names

Aberley, Doug, 228n6
Ackerman, Bruce, 85n7
Adorno, Theodor, 193, 205n8, 232–33, 236
Aeschylus, 81
Afrasinbi, K. L., 115n59
Afshari, R., 116n66
Ahmad, Nasir, 112n19
Ahmed, S. Maqbal, 112n17, 113n19–22, 114n33
Alexander, Donald, 272n19
Alford, C., 253n22
Ames, Roger, 106
Apel, Karl-Otto, 228n1
Aristotle, 21, 85n6, 87n38, 210
Attar, Farid ud-Din, 103
Attfield, R., 171n32, 253n21

Bacon, Francis, 26, 42n9
Bakunin, Mikhail, 118
al-Balkhi, Abu Zaid ibn Sahl, 101
Ballard, B. W., 202, 206n35
Barnes, Trevor, 11n1
Barrett, William, 205n13
Barry, Brian, 67, 83
Basso, Keith, 224–25
Bauman, Zygmut, 207n44
Baxter, William, 213–14
Bayly, C., 112n16
Bear, Luther Standing, 273n30

Beck, Myrl E., 136n7
Beck, Ulrich, 10, 232–33, 242–48, 252n16
Becker, L. C., 110n2
Benhabib, S., 253n23
Bentham, Jeremy, 85n14, 133, 153
Berg, Peter, 258, 272n15
Berger, Peter, 205n14
Berkes, F., 161, 170n7
Berkland, James O., 136n7
Berling, A., 115n54
Bernstein, J. M., 252n10
Berry, Thomas, 119, 126, 136n8
Berry, Wendell, 87n33, 228n8
Berthold-Bond, Daniel, 202–3, 206n33
Biehl, Janet, 260, 265
al-Biruni, Abu Raihan, 103–5, 114n46
Blackstone, William T., 271n4
Blaug, M., 170n16
Blumer, Herbert, 198
Bodin, Jean, 44n9
Boilot, D. J., 114n44
Bookchin, Murray, 122, 128–29, 135n2, 235
Bookhardt, D. E., 139n34
Booth, Annie L., 10, 273n30
Borgmann, Albert, 11n1
Botterweg, T. H., 157
Bottomore, Tom, 251n1
Boulding, Elise, 108

Boulding, K. E., 170n9
Bradley, A. C., 86n20
Breitbart, Myrna, 139n32–33
Brendan, C. C., 163
Brennan, Andrew, 253n21
Broek, Jan O. M., 42n5
Brokenna, Don, 271n2
Brown, D. A., 116n62
Brown, Joseph Epes, 269
Buck-Morss, Susan, 251n1
Burch, Robert, 6
Buttimer, Anne, 11n1, 115n53
Butzer, K. W., 112n16, 113n19

Cafard, Max, 139n31
Cahen, Harley, 187n20, 187n23
Calhoun, Craig, 253n29
Callewaert, J. H., 116n64
Callicott, J. Baird, 88n58, 106, 211, 220–21
Carlson, Rachel, 55n88
Casey, Edward, 11n1
Cassier, Ernst, 42n7
Chambers, S., 251n2
Chase, Steve, 141n54
Cheney, Jim, 225, 230n35
Chichilnisky, G., 149
Cienfuegos, Paul, 273n21
Clark, C., 170n7, 170n11
Clark, John, 8, 135n1–2, 137n10
Clarke, C., 114n49
Clayton, Anthony, 8, 161, 163, 170n19
Code, Lorraine, 226
Coleman, John, 262
Collins, Joseph, 142n72
Collins, Randall, 193
Common, M., 146, 154, 159–60, 169n1
Comstock, Gary, 221–22
Conkin, Paul, 72
Cosgrove, Denis, 11n1
Costanza, Robert, 88n56, 170n7–8
Craig, P., 169n2
Cronan, William, 228n9
Crone, P., 112n15

Daly, Herman E., 157, 170n6
Dawidowicz, Lucy S., 60n1
d'Eaubonne, Françoise, 259
Deen, M., 115n59
Dejongh, Thérèse, 138n24
Denison, Mark S., 85n4
Derrida, Jacques, 239
Descartes, René, 32
de-Shalit, Avner, 61n11
Devall, Bill, 186n2, 228n3, 229n24
Dewey, John, 196
Dols, M. W., 113n31
Dryzeck, John, 240
Dunbar, Gary, 118, 137n13
Duncan, Colin, 55
Duncan, James, 11n1
Dworkin, Ronald, 66

Eckersley, Robyn, 223, 241
Eisler, Riane, 267
Elder, J. C., 115n60
Ely, James W., Jr., 86n24
Engel, J. R., 116n64
Entrikin, J. Nicholas, 11n1
Epstein, Richard, 70
Esposito, Joseph, 111n13

Faber, M., 149, 172n49
Fackenheim, E. L., 47n86–87
Fakhry, M., 113n26, 113n28
Farber, Stephen, 88n56, 254n48
al-Faruqi Ismáil Ragi, 114n38
Feenberg, Andrew, 234, 246, 251n1
Fichte, Johann, 36–37, 40–41, 46n69
Folke, C., 161, 170n7
Foucault, Michel, 239
Fox, Warwick, 238
Foy, G., 151
Fraser, Nancy, 253n29
Freeman, A. M., 172n41, 172n51, 172n54, 252n17
French, William, 220
Freund, Ernst, 84n2

Gale, Stephen, 4–5
Gandy, Matthew, 9

Index of Names

Gedan, Paul, 42n7
Giblin, Béatrice, 119
Gilbert, Martin, 60n1
Gilligan, Carol, 142n78
Glacken, Clarence, 42n1
Godlewska, A., 110n4, 112n16, 115n51
Gole, S., 113n23
Goodman, L., 114n40
Gordon, David, 228n6
Gorz, André, 235
Gregory, D., 112n16, 253n29
Griscom, Joan L., 260
Grube, E., 113n32
Guha, Ramachandra, 222
Guttmann, Amy, 86n17

Habib, I., 112n17
Habermas, Jürgen, 10, 11n4, 223, 235–42
Haddad, Y. Y., 111n9
Hallen, Patsy, 260
Halliday, Fred, 111n11, 112n16
Hamed, Safei el-Deen, 107–8
Hamilton, Walton H., 86n24
Hanley, N., 158, 172n40
Haraway, Donna, 213, 223
Hargrove, Eugene, 74, 86n24, 107, 168, 178–79, 186, 187n24
Harries, Karsten, 196, 206n34
Hartwick, J. M., 170n12
Harvey, David, 11n1, 11n3, 191, 193, 201–2, 207n43
Hayward, Tim, 241, 253n20
Heal, G., 149
Hegel, G. W. F., 47n83–84, 81–82, 137n10
Heidegger, Martin, 9, 18, 33, 47n91, 189–207, 238n39
Held, Virginia, 110n2
Heller, Agnes, 207n45
Helmy, M., 116n65
Helsel, S., 254n44
Henderson, Hazel, 260
Hettner, Alfred, 42n6

Hirsch, Eric, 230n39
Hobbes, Thomas, 43n9
Holland, A., 154, 169n2
Honneth, Axel, 252n11
Honoré, A. M., 85n7
Hooker, M., 114n50
Horkheimer, Max, 193, 232–33
Hough, Charles, 228n6
House, Donna, 273n28
Hovannisian, R. G., 113n26
Hughs, J. Donald, 271n2
Huntington, Ellsworth, 125
Hurley, J. B., 111n12
Husserl, Edmund, 24

Iqbal, Allama Muhammad, 100

Jackson, Peter, 10n1
Jacobs, M., 170n7
James, Preston, 18
James, William, 196
Jay, Martin, 251n1
Jefferson, Thomas, 74, 87n45
Jevons, W. S., 170n18
Joas, H., 252n11
Jonas, Hans, 26

Kader, A., 111n6
Kaiser, Rudolf, 273n30
Kant, Immanuel, 7, 15–48, 191
Karamustafa, A. T., 112n17, 113n19, 113n21
Karanikas, Alexander, 87n34
Katz, C., 116n63
Katz, Eric, 7, 60n3, 60n5, 61n9–10, 61n12, 211–12, 215
Katz, Stanley, 87n45
Kay, J., 115n54
Keith, Michael, 11n3
Kellner, Douglas, 251n1, 251n5, 254n53
Kemp, R., 252n12
Kennedy, Edward S., 114n44
Keynes, J. M., 149
Khalid, F., 116n65
Khan, Hasan Uddin, 110n4

Khan, Sayyid Ahmed, 100
Kheel, Marti, 216, 227
al-Kindi, Yacub, 101
King, D. A., 113n23
King, Roger, 9, 230n44, 230n46
King, Ynestra, 265, 272n10
Klassen, L., 157
Kovel, Joel, 141n57
Kramer, G., 115n61
Kramers, J. H., 113n19–20, 114n35
Krieger, Martin, 52–53
Kroeber, Alfred, 272n7
Kropotkin, Peter, 118, 128, 138n23, 139n32
Krupnick, A. J., 172n51
Kvaløy, Sigmund, 6

Lacoste, Yves, 120
LaDuke, Winona, 268, 270, 271n3
Ladurie, LeRoy, 139n34
Lame Deer, John (Fire), 271n3
Landau, Ronnie S., 60n1
Lappé, Frances Moore, 142n72
Latour, Bruno, 239
Lefebvre, Henri, 11n1, 201
Lehr, J. H., 171n20
Leiss, William, 246–47
Leopold, Aldo, 7, 76–78, 153, 180, 182–83, 220–21, 255
Levi, Primo, 59
Lewis, B., 112n15
Lewis, C. S., 54
Ley, David, 11n1
Livingston, John, 271n2
Livingstone, Clarence, 42n2
Locke, John, 42n9, 68–74, 176
Lovelock, James, 138n23
Luckmann, Thomas, 205n14
Lukács, Georg, 205n8
Luke, Timothy W., 252n18
Lyotard, Jean-François, 42n3

MacIntyre, Alasdair, 187n17
MacPherson, C. B., 70

Makdisi, G., 112n15
Marcuse, Herbert, 205n9, 233–35, 254n52
Marguet, Y., 114n36, 114n48
Martin, Camille, 135n1
Martin, Paul S., 271n2
Marx, Karl, 137n10, 191–93, 251n8
Maser, Chris, 52–53
Masri, A., 116n65
May, J. A., 42n5
McCarthy, Thomas, 253n22
McDermott, John, 11n1
McKibben, Bill, 228n9
McLuhan, T. C., 273n30
Mead, George Herbert, 9, 189–207
Meadows, D. H., 170n4
Meinig, D. W., 230n39
Merchant, Carolyn, 218, 273n24
Midgely, Mary, 221
Mikesell, M. W., 93
Miller, David, 67
Mitman, Greg, 88n54
Montesquieu, Baron de, 125
Munro, A., 158
Murphy, A. B., 93
Murphy, T. M., III, 251n2

Naess, Arne, 140n54, 238
Nagel, E., 187n21
Naim, A., 116n66
Nash, Roderick, 187n10, 187n12
Nasr, Seyyed Hossein, 107, 114n39, 114n42, 114n45, 115n59
Nast, H., 112n16
Nedelsky, Jennifer, 73
Neihardt, John G., 271n3
Newkirk, Alan Van, 258
Nietzsche, Friedrich, 134, 239
Norton, Bryan, 5–6, 187n6, 187n16, 215–17, 228n8, 253n21
Nozick, Robert, 87n29
Nussbaum, Martha, 12n8, 81–83

O'Brian, J. O., 116n65
Oechsli, Laruen, 215

Index of Names

Ollson, Gunnar, 4–5
O'Neil, John, 253n24
Outhwaite, W., 252n9, 252n13, 253n29

Paden, Roger, 8–9
Park, C., 115n53
Parsons, James J., 272n14
Passmore, J., 153, 187n6
Paul, Ellen Franken, 75, 85n8
Paulson, Friedrich, 44n13
Pearce, D., 154–55, 157, 170n7, 172n41
Peirce, Charles W., 204n1
Penrose, Jan, 10n1
Perrings, C., 146, 154, 159–60, 169n1
Pickles, John, 11n1
Pile, Steve, 11n3
Pinchot, Gifford, 177
Posner, Richard A., 89n62
Poston, L., 111n9

Radin, Margaret Jane, 68, 70–71, 82, 85n13, 90n84
Ransom, John Crowe, 72
Rawls, John, 83, 85n6, 85n14, 86n16, 90n83
Reclus, Elisée, 8, 117–142
Redclift, M., 144
Reeve, Andrew, 87n39
Regan, Tom, 211
Rockefeller, S. C., 115n60
Rodman, J., 171n31
Rogers, J. M., 113n21
Rolston, Holmes, III, 60n9, 89n58, 187n12, 187n18–19
Rorty, Richard, 12n8
Rose, Carol M., 89n74
Rosenberg, Alexander, 11n6
Rossiter, Margaret W., 186n3
Rousseau, Jean-Jacques, 44n9, 83, 89n66
Rustin, Michael, 245
Ryan, Alan, 87n39
Ryden, Kent, 224–25

Sagoff, Mark, 186, 253n21
Said, Edward, 112n16
Said, Hakim Muhammad, 114n47
Sale, Kirkpatrick, 258, 262
Salibian, Sunny, 262–63
Sandilands, Cate, 273n29
Sauer, Carl O., 272n7
Savitt, Steven, 12n7
Sax, Joseph, 78–79
Seamon, David, 230n39
Serafy, E. S., 146
Sessions, George, 140n54, 186n2, 228n3, 229n24
Sezgin, F., 112n17
Shepard, Paul, 212
Siddiqui, A. H., 111n12, 113n19
Simpson, I. A., 171n27, 172n50, 172n55
Singer, Peter, 60n7, 187n18
Smith, David, 11n1
Smith, J. I., 115n61
Smith, Neil, 11n1, 110n4, 115n51, 253n26, 253n30
Snyder, Gary, 139n31
Soja, Edward, 11n1
Solow, R. M., 146
Soper, Kate, 252n15, 253n21
Sophocles, 81
Spash, Clive, 8, 169n2, 171n27, 172n40, 172n42, 172n50, 172n55
Spretnak, Charlene, 260
Stamp, J. Dudley, 89n59
Steelwater, Eliza, 9
Stevens, T. H., 172n40
Stirling, A., 150
Stone, C. L., 111n10
Stone, Christopher, 215
Swimme, Brian, 119
Sylvan, Richard, 13n13

Tatham, George, 42n5
Taylor, Griffith, 42n5
Taylor, Paul, 218–20, 253n21
Templet, P. H., 254n48
Tetsuro, Watsuji, 47n90
Thompson, Paul, 221–22
Tibbetts, G. R., 113n19–20, 114n34–35

Index of Names

Tibbetts, Paul, 198
Tibi, B., 116n66
Tietenberg, T., 170n7
Toelken, Burre, 269–70
Trachtenberg, Zev, 7, 90n78
Troll, C. W., 113n29
Tuan, Yi Fu, 11n1, 115n54–55, 230n39
Turner, Frederick Jackson, 176, 186n3
Turner, R. K., 154–55

Udall, Stewart, 257

Van Pelt, Robert Jan, 51–52
Vico, Giambattista, 43n9
Victor, P. A., 158–59, 170n5, 170n14

Walters, C., 150, 172n56
Walzer, Michael, 65–66, 80, 85n12
Walzer, R., 113n25
Warf, Barney, 206n39
Warnke, Georgia, 66

Warren, Karen, 223–27, 229n11, 265, 272n10
Watson, A., 114n50
Watt, W. M., 112n15
Wescoat, James, 7, 110n4, 111n6, 111n12, 112n16, 112n18, 113n21, 115n52
White, Lynn Jr., 106
White, S. K., 253n18
Wiggerhaus, Rolf, 251n1
Williams, Rosalind, 273n22
Winner, Langdon, 210
Wong, P., 110n2
Worster, Donald, 138n19
Wright, Larry, 187n22

Yahil, Leni, 60n1
Young, Don, 74
Young, Iris Marion, 11n3, 253n29

Zimmerman, Francis, 114n46
Zimmerman, Michael, 206n36

About the Contributors

Annie L. Booth is assistant professor of environmental studies at the University of Northern British Columbia.
Robert Burch is associate professor of philosophy the University of Alberta.
John Clark is professor of philosophy at Loyola University.
Anthony M. H. Clayton is a faculty member at the Institute for Policy Analysis and Development at the University of Edinburgh.
Matthew Gandy is lecturer in geography in the School of European Studies at the University of Sussex.
Eric Katz is an associate professor of philosophy, and director of the Science, Technology and Society Program at the New Jersey Institute of Technology.
Roger King is assistant professor of philosophy at the University of Maine.
Roger Paden is associate professor of philosophy at George Mason University.
Clive L. Spash is a faculty member in the Department of Land Economy at the University of Cambridge.
Eliza Steelwater is assistant professor in the Department of Urban and Regional Planning at the University of Illinois at Urbana-Champagne.
Zev Trachtenberg is assistant professor of philosophy at the University of Oklahoma.
James L. Wescoat, Jr., is associate professor of geography at the University of Colorado.

Philosophy and Geography Style and Submission Guide

Philosophy and Geography is a peer reviewed annual, with each volume focusing on a specific theme. Each issue addresses a topic of mutual interest to philosophers and geographers. The annual is edited by Andrew Light and Jonathan M. Smith, in consultation with the editorial board, and published by Rowman & Littlefield Publishers, Inc. All material submitted to the editors is subjected to peer review by members of the editorial board, associate editors of the journal, or others, serving at the behest of the editors. (Themes of upcoming issues are listed at the front of this volume.)

Length

Authors should aim for manuscripts of about 10,000 words, including notes. Shorter and longer manuscripts will be considered, but only extraordinary circumstances will justify acceptance of manuscripts of less than 6,000 or more than 12,000 words. As this length includes the notes, authors are urged to limit notes to citation of works directly relevant to their argument.

Format

Authors should send three copies of their manuscript to one of the editors.

>   Andrew Light   Jonathan M. Smith
>   Department of Philosophy   Department of Geography
>   University of Montana   Texas A&M University
>   Missoula, MT 59812-1038   College Station, TX 77843-3147

Each copy must be single-sided, double-spaced, in a large type size, with wide margins (one inch margins preferred). Illustrations submitted with the final draft of accepted manuscripts must be camera ready. Please do not send bound manuscripts. Once an article has been accepted authors must submit a disk copy of their manuscript.

Notation Style

Authors should follow the *Chicago Manual of Style* and use American spelling. Notes will be printed as endnotes and should be used judiciously. Do not use more than one note in a single sentence, and whenever possible group all of the citations and asides from an entire paragraph in a single note. Examples of some common citations:

Book: Clarence J. Glacken, *Traces on the Rhodian Shore* (Berkeley and Los Angeles: University of California Press, 1967), xiii.

Volume chapter: Roger J. H. King, "Relativism and Moral Critique," in *The*

	American Constitutional Experiment, ed. David M. Speak and Creighton Peden (Lewiston, N.Y.: The Edwin Mellen Press, 1991), 145–64.
Journal:	Eric Katz, "The Call of the Wild: The Struggle Against Domination and the Technological Fix," *Environmental Ethics* 14, no. 3 (1992): 271.
Newspaper:	Timothy Egan, "Unlikely Alliances Attack Property Rights Measures," *New York Times,* 15 May 1995, A1.

The full citation should be given only in the first note in which a work is cited. A short form should be used in all subsequent notes. Short forms of the works cited above might appear as follows:

Book:	Glacken, *Traces on the Rhodian Shore,* 372.
Volume chapter:	King, "Relativism and Moral Critique," 146.
Journal:	Katz, "The Call of the Wild," 272
Newspaper:	Egan, "Unlikely Alliances."

Originality

Authors are asked to submit a letter with their manuscript stating that the material in the manuscript has not been published elsewhere, that it is not presently under consideration by another publication, and that it will not be submitted for consideration by another publication until the author has been notified of the final decision of the editors of *Philosophy and Geography.* Upon publication, Rowman & Littlefield will possess the copyright.

Editorial Policy

Philosophy and Geography is on an accelerated production schedule. Papers accepted in February appear in print in the fall of the same year (paperback in October, hard cover in November), and the editorial work occurs in a much shorter period. It may not be possible to return either the final copyedited manuscript or the page proofs to the authors for approval. The editors will contact an author if there appears to be a need for an extensive, significant, or objectionable change to his or her manuscript, but they will not seek an author's consent to make minor alterations, additions, or deletions. These decisions are made at the editor's discretion.